"Fearing the state of being 'lost in the world we have made,' Williston roams far and wide for reference points in a time of bewildering climatic upheaval. With grand, Harari-like sweeps, this insightful romp through philosophy, literature, ecology, and technology displays the creative boldness the times demand."

—**Christopher J. Preston**, University of Montana, Missoula. Author of *The Synthetic Age: Outdesigning Evolution, Resurrecting Species and Reengineering our World*

"An accessible and engaging analysis of the ways in which the climate crisis is analogous to other, historically significant 'traumas.' This is a vitally important topic, and I applaud Williston for his creative approach to bringing its philosophical aspects to a broad readership."

—**Steven Nadler**, William H. Hay II Professor of Philosophy, University of Wisconsin-Madison. Author of *Think Least of Death: Spinoza on How to Live and How to Die*

"This lucid analysis of the crisis in Western thinking generated by climate change shows how previous historical disruptions have led to the kind of innovations in thought that we now urgently need. It should be read carefully by anyone wondering how to think and act in our new Anthropocene circumstances."

—**Simon Dalby**, Balsillie School of International Affairs. Author of *Anthropocene Geopolitics: Globalization, Security, Sustainability*

"A timely, accessible, smart, and informed discussion of the climate crisis, and our disorienting exit from the Holocene. Williston shows why philosophy matters in these times, how it can be done with passion and rigour, and what wisdom looks like for all of us worried about the future of life."

—**Todd Dufresne**, Lakehead University. Author of *The Democracy of Suffering: Life on the Edge of Catastrophe, Philosophy in the Anthropocene*

PHILOSOPHY AND THE CLIMATE CRISIS

This book explores how the history of philosophy can orient us to the new reality brought on by the climate crisis.

If we understand the climate crisis as a deeply existential one, it can help to examine the way past philosophers responded to similar crises in their times. This book explores five past crises, each involving a unique form of collective trauma. These events—war, occupation, exile, scientific revolution and political revolution—inspired the philosophers to remake the whole world in thought, to construct a metaphysics. Williston distills a key intellectual innovation from each metaphysical system:

- That political power must be constrained by knowledge of the climate system (Plato)
- That ethical and political reasoning must be informed by care or love of the ecological whole (Augustine)
- That we must enhance the design of the technosphere (Descartes)
- That we must conceive the Earth as an internally complex system (Spinoza)
- And that we must grant rights to anyone or anything—ultimately the Earth system itself—whose vital interests are threatened by the effects of climate change (Hegel).

Philosophy and the Climate Crisis will be of great interest to students and scholars of climate change, environmental philosophy and ethics and the environmental humanities.

Byron Williston is Professor in the Department of Philosophy at Wilfrid Laurier University, Canada. He is the author of *The Ethics of Climate Change: An Introduction* (Routledge, 2018).

Routledge Environmental Ethics
Series Editor: Benjamin Hale
University of Colorado, Boulder

The Routledge Environmental Ethics series aims to gather novel work on questions that fall at the intersection of the normative and the practical, with an eye toward conceptual issues that bear on environmental policy and environmental science. Recognizing the growing need for input from academic philosophers and political theorists in the broader environmental discourse, but also acknowledging that moral responsibilities for environmental alteration cannot be understood without rooting themselves in the practical and descriptive details, this series aims to unify contributions from within the environmental literature.

Books in this series can cover topics in a range of environmental contexts, including individual responsibility for climate change, conceptual matters affecting climate policy, the moral underpinnings of endangered species protection, complications facing wildlife management, the nature of extinction, the ethics of reintroduction and assisted migration, reparative responsibilities to restore, among many others.

Climate Justice and Non-State Actors
Corporations, Regions, Cities, and Individuals
Edited by Jeremy Moss and Lachlan Umbers

Philosophy in the American West
A Geography of Thought
Edited by Josh Hates, Gerard Kuperus and Brian Treanor

Philosophy and the Climate Crisis
How the Past Can Save the Present
Byron Williston

For more information on the series, please visit: www.routledge.com/Routledge-Environmental-Ethics/book-series/ENVE

PHILOSOPHY AND THE CLIMATE CRISIS

How the Past Can Save the Present

Byron Williston

LONDON AND NEW YORK

First published 2021
by Routledge
2 Park Square, Milton Park, Abingdon, Oxon OX14 4RN

and by Routledge
52 Vanderbilt Avenue, New York, NY 10017

Routledge is an imprint of the Taylor & Francis Group, an informa business

British Library Cataloguing-in-Publication Data
A catalogue record for this book is available from the British Library

Library of Congress Cataloging-in-Publication Data
Names: Williston, Byron, 1965– author.
Title: Philosophy and the climate crisis : how the past can save the present /
   Byron Williston.
Description: Abingdon, Oxon ; New York, NY : Routledge, 2020. |
   Series: Routledge environmental ethics | Includes bibliographical
   references and index.
Identifiers: LCCN 2020019790 (print) | LCCN 2020019791 (ebook) |
   ISBN 9780367506797 (hardback) | ISBN 9780367506803 (paperback) |
   ISBN 9781003050766 (ebook)
Subjects: LCSH: Philosophy and science. | Philosophy—History. |
   Climatic changes.
Classification: LCC B67 .W545 2020 (print) | LCC B67 (ebook) |
   DDC 190—dc23
LC record available at https://lccn.loc.gov/2020019790
LC ebook record available at https://lccn.loc.gov/2020019791

ISBN: 978-0-367-50679-7 (hbk)
ISBN: 978-0-367-50680-3 (pbk)
ISBN: 978-1-003-05076-6 (ebk)

Typeset in Bembo
by Apex CoVantage, LLC

CONTENTS

PART 3
Reorientation 167

ACKNOWLEDGMENTS

This book is the product of years of philosophical discussions with colleagues and students on the topics of both climate change and the history of modern philosophy. In particular I want to thank, in no particular order, Simon Dalby, Gary Foster, Christopher Preston, Steven Nadler, Todd Dufresne, Jennifer Welchman, Allen Habib, C. Tyler DesRoches, Matthias Fritsch, Frank Jankunis and Rocky Jacobsen. Audiences in numerous locations have also given me copious food for thought on these themes, and I thank all of those folks for coming out to hear me speak. A special thanks is due to my Routledge editors, Annabelle Harris and Matthew Shobrook, for seeing the value in the project and shepherding it so professionally through the publication process. Later stages of work on this book were completed while in virtual lockdown because of COVID-19. My family therefore deserves a special shout-out for allowing me to retreat to the upstairs office for whole days, and for maintaining relative household peace all the while. Work on the manuscript was facilitated by a Social Sciences and Humanities Research Council of Canada Insight Development Grant for which I am very grateful.

INTRODUCTION

The issue

In early 2020, COVID-19 burst upon the world, killing huge numbers of people and sending the global economy into a tailspin. This is, indisputably, a *crisis* for humanity, almost a textbook definition of what we mean when we use that word. Even as many countries began to 'flatten the curve' of infections and deaths heading into the summer, worries emerged that we would experience not a single peak-and-decline but a sine wave extending perhaps for two years (or until a vaccine is discovered). We should reflect on that image of a sine wave, a continuous series of disasters, because it captures perfectly the more significant emergency behind the headlines about the virus: climate change. In the case of climate change the wave is going to roll along for decades and centuries. While in no way diminishing the unique horrors and challenges of the novel corona virus, we should also see its arrival and spread as an adumbration of this bigger event. Adopting this perspective might allow us to prepare—socially, politically, emotionally, existentially—for a long future of similar crises. This book is meant to help with that preparation.

In both popular and academic discourse, 'crisis' often travels with two other weighty C-words: civilization and collapse. Where the crisis is perceived to be big enough, these other two C's are never far behind. We are regularly informed that if the climate crisis deepens collapse looms, and that *this* represents a threat to everything we hold dear, i.e., to civilization. It's pretty easy to see how this kind of talk can lead to dangerous hyperbole. Nothing seems to get people as worked up as what they take to be a threat to civilization. It conjures up images of once taken for granted truths and values being cast aside, often entering the picture via the hordes outside (or frankly, inside) our borders who do not share our values and are therefore the enemies of civilization.

And yet, notwithstanding the potential for hyperbole, we clearly *are* in the midst of a self-imposed ecological crisis, one so dire it challenges us to reconceive just about everything we think we know about the career of our species on this planet. Climate change is in the process of reconfiguring everything. Not just material things like our energy infrastructure but also less tangible things like our geopolitics and our values. Because its effects are going to increase in intensity in the coming decades, this transformation is going to be felt with increasing urgency everywhere in the world. None of us will escape the challenges it presents, even though some of us—the global South in particular—will bear the brunt of them more squarely than others.

How should we orient ourselves to this new reality? Are there examples from the past of humans engaging in the comprehensive reorganizations of their worlds that I think we are facing now? And just what good is a *philosopher* in helping us to sort all of this out? Good questions, all of them. In the 10 years or so that I've been writing, talking and teaching about climate change I have encountered such questions, or permutations of them, repeatedly.

The audiences I have spoken to and with over the years have been very diverse, from church groups to members of NGOs, seniors groups, indigenous peoples, young people and business people. Above all, I've been struck by just how *worried* people are about the climate crisis. It's the same sense of frustration and foreboding we saw playing out all over the world in September, 2019, when more than 6 million people in 185 countries took to the streets demanding that our political elites take the climate crisis more seriously.

The anecdotal evidence of widespread anxiety is borne out by recent polling results. 2019 was a watershed year in global awareness of climate change. The Yale Program on Climate Change Communication (2020) released a poll measuring attitudes towards this threat in the cradle of climate change denial, the United States. It asked subjects to rank their concern about climate change by reference to six criteria: dismissive, doubtful, disengaged, cautious, concerned and alarmed. Between 2014 and 2019, the first five categories saw a downward slide of 3–7 points. The number of those alarmed by climate change, however, rose *by 21 points.*

Beyond the US, we encounter similar numbers. According to a Pew Research poll conducted in 26 countries and published in 2019, climate change is perceived as the biggest threat to humanity. It places ahead of ISIS, cyberattacks from foreign countries, North Korea's nuclear program, the condition of the global economy, the power and influence of the US, and Russia's power and influence (Poushter and Huang, 2019).

Climate change has been at or near the top of lists like this for some time now. It will likely be overtaken by worries about COVID-19 in the months and possibly years to come. But the climate crisis is here to stay and will therefore never fall far from the top spot. So it's worth reflecting on this fact by comparing climate change with the other threats identified in the Pew survey. They are all very significant worries, of course. However, there are two important differences between all of them and climate change.

The first is that unlike climate change they are relatively ephemeral. However tricky, they are problems that we can imagine resolving in our lifetimes. This is because they are mostly confined to geopolitics. As challenging as they may be they do not force us to ask questions that transcend this relatively circumscribed sphere. The climate crisis is different. It is the diametric opposite of an ephemeral crisis. Some of the carbon we are putting into the atmosphere now will remain up there, altering the global climate, for thousands of years. Many scientists now suggest that we have effectively *cancelled the next ice age*, which had been scheduled to roll over the planet in about 50,000 years (Stager, 2011).

Climate change connects our activities not just to distant stretches of future time, but also spatially outward into the whole Earth system. In the process, it *encompasses* our geopolitics, the focal point of all those other threats. More than that, it swallows them whole. Many security experts are convinced that climate change will re-order the geopolitical map over the course of this century. Obviously, the causal arrow does not go the other way. That is, the problems of cyberattacks, North Korea and so on will not, all by themselves, affect what happens to the global climate. We can safely conclude that climate change is both the biggest crisis we face and the one that is, by far, the most comprehensive in potential scope.

The second—and related—point of difference is that, among all these threats, only the climate crisis demands big picture thinking from us and about us. None of the other threats compels us to ask where we are going *as a species*, and what it is about us that has brought us to the current impasse. I'm going to demonstrate that the climate crisis *does* demand this kind of thinking. Even better: I'm going to provide it.

What exactly is everyone so worried about? It's simple really: they are worried about whether or not the societies in which they live are well-organized enough to deal adequately with persistent catastrophe. The speed with which COVID-19 spread in any country was a reflection of how rationally that country was governed. Taiwan, New Zealand and Singapore did a pretty good job containing it, while Iran, Russia, Brazil and the US responded with varying degrees of incompetence. In the case of the US the mismanagement can be traced directly to President Trump's early efforts to soothe Americans with what became known as 'happy talk' about the threat they faced. This happened most significantly during the crucial month of February, 2020, when the disease could have been contained had robust social distancing as well as testing and contact-tracing measures been put in place.

As I have said, climate change will bring waves of crises like this. We're facing decades of compounding and cascading disasters, from wildfires and floods to uncontrolled migrations, droughts and the spread of more deadly diseases. One mega-crisis is difficult enough, but serial crises on this scale will challenge the coping abilities of even the wealthiest and most resilient societies (Dufresne, 2019). Serial crises will therefore produce severe material deprivation and resource scarcity, including scarcity of health care resources. Once that phenomenon kicks in, ever-widening cracks in the edifice of civilization will appear. I sincerely hope that COVID-19 is causing people to wake up to this new reality.

People are therefore right to be worried. I want to emphasize that. Crisis is here to stay for the foreseeable future. This is not a book about what life will look like on the other side of crisis. It is not a 'beyond crisis' narrative. It indulges no utopian dreams about a future in which humanity has cast aside the yoke of angst-ridden adaptation to disaster. Let others write those books. In their day, Marx and Engels were fierce critics of what they called utopian socialism, something of a trend among 18th–19th-century social thinkers like Charles Fourier, Robert Owen and Henri de Saint-Simon. The following analysis is not Marxist in any meaningful sense, but I do take this anti-utopian animus to heart. We must act as though crisis is here to stay and focus on *saving* what is truly valuable in our world. Hence this book's title.

It's nothing short of amazing that a 200,000-year-old species has rushed into an existential crisis of this magnitude over the course of just 150 years of industrialization. From the standpoint of any period in our history, the pace of development over this period has been entirely off the scales. Because the unforeseen negative consequences of this development are so far-reaching—temporally and spatially, politically and ethically—piecemeal analyses of our predicament are no longer going to teach us very much about what we have done and what is at stake. Our crisis, I'm going to suggest, is fundamentally about the loss of an existential home—the Earth as we have known it over the long course of the Holocene epoch—and the cosmic disorientation this anthropogenic expulsion has brought in its train. Though we have various more or less clever ways of covering over the fact, we are, in short, a species in free-fall: directionless, panicky and reeling. To grasp this predicament we need the big picture. We need philosophy.

The apology

Given the peculiar focus just outlined, this book is likely quite a bit different than anything else you have read about the climate crisis. As a philosopher I take to heart the Socratic notion that wisdom requires self-knowledge. So this is not a book about green policy or what steps you can take to reduce your carbon footprint. Those are important issues, and I hope that what I have to say here inspires you to take them more seriously than you might have so far, but those would be secondary benefits of my analysis. My main aim is to enhance your knowledge of who and what we are at this perilous moment in our history.

As I see it, arriving at such collective self-understanding is an inherently *metaphysical* task, and that's the bit that almost everybody writing and talking about the climate crisis has overlooked. What is metaphysics? It is the study of reality's fundamental structure. Metaphysics asks questions about God, the nature of mind, the number of basic elements, or 'substances,' in the world, the relation between the human and the non-human parts of reality, and more.

A metaphysical *system* is an account of how all these phenomena—Nature, God, mind, matter, time and history, politics and even technology—fit together. American philosopher Wilfrid Sellars (1912–1989) once said that metaphysics (or philosophy more generally) is concerned with "how things in the broadest possible sense of the term hang together in the broadest possible sense of the term." Metaphysics is world-constitution in thought.

I'm going to show that this constructive task was rarely perceived by its most significant practitioners as an end in itself, a mere intellectual game. Rather, it almost invariably happened as a response to perceived crisis. Moreover, the shape taken by the crisis is in some sense less important than the perception by a certain collective that it threatens a way of life, a concrete way of being in the world, the values that structure collective existence.

The philosophers apprehend a world, their world, in deep crisis of one form or another and diagnosis it as a flaw in how reality's fundamental structure has been understood so far. Their job as they see it is thus to reconstitute the world at this level. I will spend most of the book showing how certain canonical philosophers have done just this in the face of the crises that shook their worlds, and that these examples can help us understand our own impasse. The enterprise might strike you as hopelessly quixotic. How could a discipline as unapologetically abstract and persistently technical as philosophy be applicable to something as concrete as the climate crisis?

To buy into the program I'm about to present, you need to believe that ideas matter. By that I mean that if you think that ideas, including those of philosophers, are historically impotent gas-clouds floating around in the cultural ether, then this book is not for you. I don't think many people are like this, so the ones that are don't worry me too much. But I *do* worry about peoples' perception that philosophical ideas are beyond their grasp. I heard a radio program recently about a philosopher who had set up a little booth in the New York subway, with a banner reading "Ask a Philosopher." Anybody could walk up to the guy and ask him a question, like "how can we know that we are real?" This led to some charming and insightful exchanges.

I applaud that philosopher for making the discipline accessible in this fashion. But I was dismayed by the journalist's first words to him: "Most of us find philosophers incredibly *daunting*!" Trust me, I get it. Too much professional philosophy *is* narrowly focused, over-technical and thoroughly jargon-saturated. But philosophy need not be presented this way, and for most of its history it was not.

Almost every professional philosopher you are likely to meet got into the discipline by being roused, usually at some point in their late teens or early 20s, by meaning-of-life questions. And then they were compelled to make their stolid way through a system of professional training whose chief aim seems to be to smother every vestige of this original enthusiasm. The problem shows up even in areas of philosophy that you'd think might be immune to it. In a nuanced study of postwar liberal political philosophy the Harvard Professor of Government Katrina Forrester laments that only philosophers could, for example, have turned the vital issue of ecological survival that emerged in the 1970s into "an anodyne puzzle" (2019, 174).

But batter them as the experts will, the questions themselves never go away. Have you ever wondered whether or not you are truly free? Whether or not there is a God? If the human mind is just a hunk of matter or, perhaps, is something non-material? If there's a heaven, or some other 'place' we might go after this life? Where the distinction between right and wrong comes from? Whether you might, right now, be in the matrix or if robots deserve rights? If you have pondered any of these questions explicitly, you have been doing philosophy.

For that matter, if your imagination has been sparked about questions *like* this by watching, say, *The Good Place*, *The Matrix* or certain episodes of *Black Mirror*, then you have also been doing philosophy. Philosophical questions arise all the time, entirely spontaneously. They are natural and irrepressible. Considered as a *discipline*, Philosophy simply insists that we think about such problems systematically, with as much logical rigor and clarity as we can muster. There's no special technique to learn beyond this rigor and clarity.

So this is not a book primarily for professional philosophers. I do not attempt to survey and synthesize the vast amount of philosophical literature that has accreted around, say, Descartes or Hegel over the ages. Specialists on these and other philosophers—or on the philosophical *problems* I talk about here, like theories of freedom, the nature of moral value, the metaphysical status of sensations and feelings, and more—will no doubt wince at the breeziness with which I sometimes treat them. Most of my own work has been in technical philosophy, so I appreciate the complaint. But in this case, it is misplaced.

At the most general level, that 2,500-year-old body of literature we refer to as The History of Western Philosophy is best characterized as a tangled skein of brilliant insight and unbelievable tedium. One way to understand what I'm up to in this book is that I'm ignoring the tedious bits and focusing on (some of) those parts of the brilliant stuff that we might use to help us grasp our current impasse more deeply and effectively.

My warrant for passing over so many argumentative details, or being so selective about the ones I do present, then, is that I'm not engaging in technical philosophy here. Rather, I am attempting to cast the climate crisis in a particular light, one inspired by the sort of big philosophical questions that have been posed by the most important thinkers in the history of the discipline. And I'm doing this explicitly *for* intelligent members of the public who happen to have minimal formal training in philosophy, or even none at all.

The argument

With that apology out of the way, here's a summary of the book's key claims. Part I lays the groundwork for the rest of the analysis by defining the form and content of the climate crisis. The main aim is to give substance to the concept of climate crisis understood as existential disorientation and homelessness. I begin

(Chapter 1) by arguing that the age's most distinctive mood is, and should be, a sense of bewilderment at the mess we have made of our ecological home. This mood, I suggest, can be a source of moral clarity. Next (Chapter 2), I show that the metaphysical source of our current malaise consists primarily in being stripped of any possible foundations for our values. The only possible candidates for such foundations are God and Nature, so I spend most of the chapter showing why appeals to them don't work now.

Bringing the insights of Chapters 1 and 2 to bear, the final chapter of Part I (Chapter 3) explains exactly why climate change is a crisis. Here, I examine the way our core values are threatened by the various forms of material scarcity that climate catastrophe is likely to induce. This is where I stake the main clam of Part I: that ongoing crisis is now effectively a part of the human condition. It is going to redefine who and what we are. If I've done my job properly in this part of the book, you will come away from it with an enhanced understanding of the profound crisis-induced disorientation of our time, and hopefully also *feel* that disorientation in your bones.

That brings us to Part II, the book's core. The five past crises I have isolated challenged a group's self-understanding root and branch. Each involved a unique form of collective trauma and drew forth a response from a famous philosopher: the Peloponnesian war in ancient Greece, as experienced by Plato (Chapter 4); invasion and occupation in the late Roman Empire, as experienced by St. Augustine (Chapter 5); the modern scientific revolution, as experienced by Descartes (Chapter 6); the persecution and exile of the Sephardic Jews in the 16th–17th-centuries, as experienced by Spinoza (Chapter 7); and the French Revolution, as experienced by Hegel (Chapter 8). In the hands of the philosophers the traumas result in a metaphysics, from each of which I extract an intellectual innovation that can help reorient us in the midst of our crisis.

Those five innovations are: that political power must be constrained by the knowledge of climate scientists, that is, that our democracy must also be a rule of the knowers, an 'epistocracy' (Plato); that such power must also be constrained by care or love of the ecological whole of which we are inevitable parts (St. Augustine); that we must not shrink from the job of enhancing the design of the 'technosphere' (Descartes); that we must learn to see the whole Earth as a system and act to preserve its internal complexity (Spinoza); and that we must grant *rights* to anyone or anything—ultimately the Earth system itself—whose vital interests are threatened by the effects of climate change (Hegel).

In drawing connections between past collective traumas, the construction of metaphysical systems and the uses we can make of all this for our own purposes, I have tried to avoid two interpretive pitfalls. First, I am decidedly *not* claiming, for any of these philosophers, that there is a deterministic relation between the collective trauma and the philosophy. None of these philosophers were compelled by some mysterious force to produce any philosophical response to the crisis at all; and all of them could have come up with a different metaphysics, one that had little or nothing to do with the collective trauma specific to their times.

Second, I don't claim there's no value whatsoever in studying metaphysics apart from the uses we can make of it to confront our own historically specific social and political challenges. Commentary on the great thinkers in the Western canon has often uncovered fascinating insights into the mind, God, history, etc. with no attempt to make these problems and puzzles particularly relevant to the commentators' own social world. I have no problem with this interpretive approach, having engaged in plenty of it myself over the years. But not here.

Resolving to avoid these two errors makes it easier to say something more positive about my methodology. It is this. In my view, metaphysical systems can become far more interesting once we connect them up with social reality the way I do here. Establishing this connection allows us to appreciate aspects of a philosopher's worldview that may have looked insignificant or implausible absent the connection. And then again, discovering that previous thinkers have tried to reorient their worlds in the face of crisis can make those thinkers, and those worlds, seem a little less strange to us than they might otherwise have seemed. This can give us courage in the face of our own crises by building bonds of imagination among humans across the centuries.

Part II provides essential content to the metaphysics we need now. But it is not sufficient. We must also craft a distinctive metaphysics for the age of climate crisis, which I do in the book's final Part (Chapter 9). As I argue, our world is now fully technologized. There is no longer a discernable or meaningful gap between our technologies and the rest of the world, from the atmosphere to the non-human biosphere to our cities and our own bodies. It's all one thing. I call this 'Anthropocene monism' and lay out two broad features of it: a new philosophy of history and a reinvigorated understanding of the political control we must exert over the technologies pervading our lives. If we manage to internalize the paradoxical demands of this metaphysics—supplemented by the lessons from Part II—it can aid us in reorienting ourselves in the tumultuous world we have made. This, in turn, might help re-energize the human enterprise.

At its best, philosophy is a universally accessible discipline. It is a portal to the deepest questions of human existence. It begins in wonder and culminates in a form of enhanced self-understanding that no other mode of thinking can match. It probably would never have arisen in the first place but for the many adversities to which we humans have always been exposed. Asked to define philosophy, the American philosopher Rebecca Goldstein—echoing Sellars—answered that "it is the attempt to get our bearings, as broadly and systematically as possible; people should study it because everybody's trying, as best they can, to get their bearings" (Goldstein, 2019). I have written this book in the belief that philosophy can help all of us get our bearings in the midst of the peculiar form of disorientation set in motion by the climate crisis.

References

Dufresne, T. (2019). *The Democracy of Suffering: Life on the Edge of Catastrophe, Philosophy in the Anthropocene.* Montréal: McGill-Queens University Press.

Forrester, K. (2019). *In the Shadow of Justice: Postwar Liberalism and the Remaking of Political Philosophy*. Oxford: Oxford University Press.

Goldstein, R. (October 10, 2019). *What Is It Like to Be a Philosopher?* Retrieved from: www.whatisitliketobeaphilosopher.com/rebecca-goldstein?rq=Rebecca%20gold stein. Accessed October 18, 2019.

Poushter, J., and Huang, C. (February 10, 2019). "Climate Change Still Seen as the Top Global Threat, But Cyberattacks a Rising Concern." *Pew Research Center.* Retrieved from: www.pewglobal.org/2019/02/10/climate-change-still-seen-as-the-top-global-threat-but-cyberattacks-a-rising-concern/. Accessed April 10, 2019.

Stager, C. (2011). *Deep Future: The Next 100,000 Years of Life on Earth.* New York: Harper-Collins.

Yale Program on Climate Change Communication. (2020). *Climate Change in the American Mind.* Retrieved from: https://climatecommunication.yale.edu/publications/climate-change-in-the-american-mind-november-2019/. Accessed January 16, 2020.

PART 1

Disorientation

1

IN PRAISE OF BEWILDERMENT

Richard III, King of England and Ireland for two tumultuous years (1483–1485), is said to have murdered his nephews because he considered them his political rivals. He died in the battle of Bosworth Field fighting Henry of Tudor, later King Henry VII. In 2012, archaeologists dug up Richard's bones in a parking lot in Leicester. CT scans of the skull revealed multiple piercings and blunt-force wounds, indicating that he had likely been hacked and stabbed to death on the battlefield by multiple attackers. Whether or not he really was a child murderer, this tyrant had a surfeit of foes and they clearly had a score to settle with him.

Elizabeth I, by contrast, ruled the realm for 45 years (1558–1603), overseeing an unprecedented flowering of British culture in these years. Her death (by cancer or blood poisoning) was evidently not pretty, but at least she did not have an entire royal House baying for her blood. All political leaders believe they act for the good of the whole order they rule. But some of them have a more justified claim to this belief than others. Though there is of course much gray area in these matters it's sometimes pretty easy to distinguish between benevolent and tyrannical monarchs.

The Book of Genesis asserts explicitly that humans were set on Earth to rule over the rest of Creation and this is exactly how we have behaved ever since that book was written: as presumed monarchs of the biosphere. Assuming we are now in a position to evaluate it critically, what should we say about our long reign? Have we behaved more like Richard or Elizabeth? Or are we a confusing and disturbing mixture of the two models of planetary governance? The answer does not point in a particularly encouraging or flattering direction.

Recently, the Intergovernmental Science-Policy Platform on Biodiversity and Ecosystem Services (IPBES) issued a report demonstrating that we are in the process of wiping 1 million species off the planet, and that the rate at which we are eviscerating the non-human world is accelerating rapidly. The group's Chair, Sir Robert Watson, summarizes its findings this way:

The overwhelming evidence of the IPBES Global Assessment, from a wide range of different fields of knowledge, presents an ominous picture. The health of ecosystems on which we and all other species depend is deteriorating more rapidly than ever. We are eroding the very foundations of our economies, livelihoods, food security, health and quality of life worldwide.

(United Nations, 2019)

This is certainly not how we had imagined things working out. The ancient Greek poet and playwright Sophocles (497–405 BCE) wrote a paean to the extraordinary capacity of humans to alter nature for the benefit of all its members. The poem praises our technological transformation of the "holy and inexhaustible" Earth, exulting in our wise and beneficent sway over everything from "lightboned birds" to the "sultry mountain bull" (Sophocles, 2007, 909). This sort of talk is fairly typical of our early self-conception as nature's benevolent monarchs, but it endures even today.

Between the paean and the report lies some 2,500 years of continuous economic development, which towards the end morphed into breakneck hyper-industrial expansion powered by fossil fuels. Both documents express a form of shock, but shock can come in at least two varieties. The first is what we call wonder. It's a mostly positive reaction to something strange or unexpected. The second is bewilderment, something scarier or at least less sure of its footing. Sophocles is expressing wonder, while the IPBES, in my view, is expressing bewilderment. Taken together, then, Sophocles (as well as the Book of Genesis) and the IPBES describe the historical arc of our relation to the non-human world over the last few millennia, an arc travelling from excitement to fear, wonder to bewilderment.

This book begins with a reflection on the bewilderment so many of us are experiencing because of climate change. Sophocles' poem rings out with confidence in the human enterprise. Although it shows up relatively late in the Holocene, this is the cocksure self-assessment of humanity typical of that geological epoch. Since that time, an interglacial period beginning roughly 12,000 years ago, we humans have made a secure home for ourselves in an otherwise indifferent world, and we have done this by subduing the rest of the biosphere. In the process, we have made the world a familiar place.

By contrast, the IPBES report is a lament for a lost home. And not just any home, but our only home, the very same home identified in the poem. We have fouled our own nest so thoroughly that only what Watson calls "fundamental, structural change" can now save it. This is how I'm going to talk about bewilderment in this chapter. It is the feeling of unfamiliarity where we expect familiarity, the sense of homelessness where there should be a home, the idea that our own cleverness and plastic adaptability have set us adrift or banished us from our primal place of origin and belonging.

Put otherwise, we have made the world *uncanny*. To call a phenomenon uncanny is to identify it as creepy or weird, but in the specific sense that it represents a strange meeting of the familiar and the unfamiliar. This creepiness usually comes with a sense of threat. Think, for instance, of humanoid robots, which inspire this feeling in many people. They are a lot like us, but there's something in their movements or speech patterns that is not *quite* right. We feel vaguely that they might turn on us at any moment, that their familiarity is really just a ruse to mask the threat coiled in them. We have the inchoate sense that they are fundamentally unpredictable and thus untrustworthy.

My task in this chapter is to convince you that this can be a fruitful way to think about life in the age of climate change, a transformative event set in the larger context of our brand new epoch, the Anthropocene. We are emerging from a period in which a stable climate has given us a very familiar world, one whose patterns and rhythms we have been allowed to take more or less for granted. What we are passing into is less clear, but it feels both weird and threatening.

There's a much wider context of instability that must be explained as well. I'm talking about the instability of our values that comes with being denizens of the modern age. The mark of the modern, as we'll see, is to be deprived of traditional sources of moral meaning without having any obvious substitutes for them. The result of this layered instability—modernity *plus* the climate crisis—is a world that is quintessentially uncanny. It is bound to bewilder us but, as I will argue, we should embrace this bewilderment as a form of existential therapy and a path to moral clarity.

Pierre and the pundits

In this part of the book I'm going to describe what it means to say that we have entered a new historical epoch or time, one marked essentially by crisis. But for the moment I'm mostly interested in how this *feels* to us. I want to investigate a particular, highly complex collective *mood*: bewilderment. To get a sense of what an analysis of this sort involves I'll begin by contrasting my approach with others who are operating in the same conceptual field.

Yuval Noah Harari's massively popular books, *Sapiens: A Brief History of Humankind* (2016) and *Homo Deus: A Brief History of Tomorrow* (2017), take us on whirlwind tours of the history of our species, all with a view to illuminating the contours of our tech-shaped world as well as the future that likely awaits us. Though he's also not a trained historian, Stephen Pinker's *Enlightenment Now: The Case for Reason, Science, Humanism and Progress* (2018) is in the same vein. The genre is proliferating (e.g., Shapiro, 2019; Cohen and Zenko, 2019). Let's label these thinkers the *pundits* of the new age. They are remarkably sanguine about our times.

Because I think it is now essential for us to see the big-picture of our species' relatively brief reign on this planet, the pundits perform an invaluable function. They are trying to enhance our historical self-understanding, and they get the story at least partly right. But it's that 'partly' that nags at me. I came away from the books just mentioned feeling enlightened but also with the sense that something important was missing from them. One review of Pinker's book notes that the view of our post-Enlightenment history Pinker gives us is from 30,000 feet (Potter, 2018).

It's a relentless compilation of data, all designed to show that with respect to key markers of well-being—education levels, health, wealth, the spread of democratic institutions, etc.—we're demonstrably better off now than at any point in the past history of our species. Pinker thinks he needs to say this because Enlightenment values of progress and reason have come under sustained attack of late, much of it taking the form of a pouty Left dystopianism.

I have no quarrel with Pinker's basic claim, and in what follows I promise not to pout. Still, something *is* clearly missing here, something connected to the head-spinning *elevation* of Pinker's analysis. Though it's a slight oversimplification, reactions to Pinker's book tend to fall into one of three camps. The first—think of Bill Gates, who *loves* Pinker's books—says, 'Yes, exactly, the Enlightenment project is sound, and we merely need to push fearlessly ahead with it to ensure an even brighter future.' The second group contains the pouters I have just mentioned, those who think the Enlightenment project is a disaster, and has been from the start, mostly because it got entangled almost from the beginning with a heartless and overreaching capitalist economic system.

But there's a third group, and this is the one that interests me the most. To get a sense of what this group is thinking, come back to that Pew survey cited in the Introduction. In all 26 countries surveyed a majority think that climate change poses the biggest threat to global stability. The average number responding this way is 69%. Perhaps most surprisingly, climate change tops the list even in places where you'd think other perceived threats would beat it. In France, which has a recent history of domestic terrorism, 83% believe that climate change is the most significant threat. In South Korea, whose citizens live with the very real threat of *nuclear annihilation* from the regime to the north, the number is 86%. Brazil is a country torn apart by inequality, extreme poverty and the rise of a quasi-fascistic political class, and yet these challenges would seem to pale in comparison to the threat of climate change, which 72% of Brazilians list as the biggest problem in the world.

By their nature, surveys cannot capture all the reasons underlying the answers people give to the questions posed to them. But think about the level of fear and trepidation regarding climate change this survey nevertheless does reveal. Assume for the sake of argument that many of the people who put climate change at the top of their list of worries also believe in progress and humanism. If this describes even half of them accurately, we're still talking about almost 35% of the people on this planet.

These folks might insist that the IPBES is on to something really important in noting that *fundamental, structural change* is required for Pinker's ideals to be fully realized. This is a possibility that never seems to occur to those sunny optimists in the first group. And so, my people—those in the third group—feel confused, threatened, creeped out and bewildered by the events they are living through. I applaud their stance and have written this book to help them achieve their own enlightenment now.

That's Pinker, but what about Harari? The two Harari blockbusters, which taken together purport to encompass the whole past and likely future of our species, run to over 900 pages. And yet they contain a total of about 10 pages on climate change. Most of this consists in a string of bromides about how high atmospheric concentrations of greenhouse gasses have become, how ineffective our political response to this has been and how most of the effects of climate change will be visited on the global poor. There's nothing incorrect about this, but it can be found in a thousand other treatments. Harari underrepresents the full scope of the threat posed by climate change precisely because he banalizes it so thoroughly. How can someone writing a 'history of tomorrow'—however 'brief'—treat what is by far the most important problem of the near and medium future in such a cursory way?

We don't need to dispense with these analyses, but they need supplementation. I want us to be much more serious about what *crisis* means than the pundits are. To do this, we need to descend from the statistical heights and put our feet on the ground. Improbable as it might at first seem, I think Tolstoy can help us get started on this. In *War and Peace*, Tolstoy is not just telling a rousing story about how the Napoleonic wars unfolded in Russia at the beginning of the 19th-century, through the eyes of a few noble families. He is also railing against the same blinkered overconfidence I see in some of the popular pundits of our day.

Tolstoy's narrative is populated by a huge array of characters, many of whom share one important trait: they think they *understand* what is happening in Europe as Napoleon's *Grande Armée* rolls ominously from West to East. Each character fastens on a pet theory of human nature or an allegedly superior knowledge of this or that military actor in order to proclaim with confidence that history's got a discernable direction. The obvious problem, ruthlessly exposed by Tolstoy's narrative, is that the proclamations contradict one another.

But there's a deeper problem. Each of these characters interprets the world at war in a way that confirms what he or she had already believed about it. It's a social world brimming with confirmation bias *avant la lettre*. As a result, nobody is ever genuinely *surprised* by what really happens, even though some pretty unexpected things transpire, like the burning of Moscow—"the sacred and ancient capital of Russia"—after the French troops enter it in early September, 1812.

The most glaring example of this tendency is Napoleon himself, whose very ordinariness in this sense would have infuriated the man himself. The moral seems to be that when it comes to large-scale historical events, like continental wars, we should be more open to admitting frankly that we don't know what the hell is happening. Still, if Tolstoy were doing nothing more than describing the fog of war, the message would be a bit underwhelming. There's much more going on in his narrative, in particular material that can help us make sense of our own representations of the climate crisis.

Let me elaborate on this by looking briefly at Tolstoy's great anti-pundit, Piotr Kirillovich Bezuhov, aka Pierre. Pierre is a dreamy sensualist, at least in his youth (he hardens up a bit later in life). Superficially, what sets him apart from the novel's endless parade of salon pundits is that he seems, most of the time, completely out-to-lunch. But this description conceals a more important difference. Although he becomes, in the course of time, one of Russia's wealthiest men, Pierre has no fixed practical identity. His defining qualities are impulsivity, passivity and moral weakness.

Pierre bobs up and down like a cork in the stream of historical events. He fights a duel, gets drunk (often), marries a deceitful woman, falls in love with his best friend's girl (whom he later marries after the friend dies in the war), befriends French officers, throttles a would-be rapist, saves a child from a burning building, joins the Freemasons, straps a bear to the back of a policeman and throws them both into the river, plots the assassination of Napoleon, spends a month in a shack as a prisoner of war and much more.

But Pierre does not exactly *do* any of these things so much as notice that they happen *to* him. He is perennially bewildered by the shape the world takes through him and around him. From the standpoint of the meaning of history, Pierre is the arch enemy of the pundits, but in spite of this—or perhaps because of it—he is also, somehow, tuned in to the real forces driving history forward. He experiences a world in tumult as . . . a world in tumult! He embodies a general mood of disorientation, rupture, uncertainty, discontinuity, uncanniness, dread and absurdity.

These are moods permeating Russian society at the time, a current running just below the surface of society's polite chatter. Pierre is the only one who does not flee them or allow them to be smothered by parlor room prattle. The beautiful irony of Pierre is that this non-agent has a more accurate view of the meaning of his times than those who see themselves as important players in history or as having special insight into its workings, all of whom—including Napoleon—are deluded to one degree or another.

Because of this, Pierre is drawn constantly to ponder the way we humans fit into the larger cosmic whole. This is what draws him into the orbit of the Freemasons. Pierre is the reluctant philosopher of *War and Peace* (as befits him, he never settles permanently on any single philosophical explanation of things). This, finally, is what I want to emphasize most about this existential anti-hero. In this character we have an intimate connection between the experience of radical disorientation and the desperate and continual search for philosophical reorientation, for an account of the cosmic whole and our place in it.

Somehow, precisely because he seems to grasp just how absurd the world has become, Pierre emerges as a man of insight and wisdom to those around him, especially the peasantry. After the war, this feeling leads him to believe that this special insight confers moral responsibilities on him:

> The very qualities that had been a hindrance, if not actually harmful, to
> him in the world he had lived in—his strength, his disdain for the comforts

of life, his absent-mindedness and simplicity—here among these people gave him almost the status of a hero. And Pierre felt that their opinion placed responsibilities on him.

(Tolstoy, 1994, 1273)

These responsibilities include political agitation on behalf of Russia's serfs. In short, by tapping into what Tolstoy calls the "mysterious force" of history, this character understands better than anyone else what it means to *live through* crisis, and what must be done politically to make the experience less horrible. Because he embraces bewilderment in the face of chaos, he finds a path to moral clarity and a sense of political responsibility. For this reason, he's an exemplar for us.

The madman

Why should we think of what is going on in our times as analogous to what Tolstoy depicts in Russia from 1800–1815? We'll come to the full explanation of that shortly, but before we do I want say more at a general level about the nature of such disorientation. If we're to employ the notion of collective moods responsibly we need to emphasize that such phenomena are always historically unique. In our own case this entails focusing in the first place on the fact that we are thoroughly *modern*. This is not meant as a compliment. It's just that, like it or not, we are the semi-distant offspring of the Age of Enlightenment, which is itself more of an ongoing and not quite coherent saga than a properly fenced-in block of time.

This statement requires a caveat, which is that the modern period itself has deep historical roots reaching at least as far back as ancient Greece and Rome. There are no clean breaks in history, including the history of ideas. So there's also no conflict between the claims (a) that we are moderns through and through, and (b) that modernity has a history that must be exposed if we are to understand it thoroughly. Why do I insist on fixing my analysis on the modern period? Because whatever else it is, this period has often been described by those philosophers who grasped its nature most deeply as one of profound crisis. If we are going to talk sensibly and informatively about *our* crisis, we had therefore better begin with these older insights.

The great 19th-century German philosopher Friedrich Nietzsche (1844–1900) is our best entry-point to the crisis of modernity, so we are going to spend some time with him here. In one of the most arresting passages in the whole Western philosophical canon, Nietzsche asks us to imagine a "madman" bursting into a marketplace and confronting the people gathered there with these words:

> The madman jumped into their midst and pierced them with his eyes. "Whither is God?" he cried; "I will tell you. *We have killed him*—you and I. All of us are his murderers. But how did we do this? How could we drink up the sea? Who gave us the sponge to wipe away the entire horizon? What were we doing when we unchained this earth from its

sun? Whither is it moving now? Whither are we moving? Away from all suns? Are we not plunging continually? Backward, sideward, forward, in all directions? Is there still any up or down? Are we not straying as through an infinite nothing? Do we not feel the breath of empty space? Has it not become colder? Is not night continually closing in on us? Do we not need to light lanterns in the morning? Do we hear nothing as yet of the noise of the gravediggers who are burying God? Do we smell nothing as yet of the divine decomposition? Gods, too, decompose. God is dead. God remains dead. And we have killed him.

(Nietzsche, 1974, 181)

This is the now classic modern statement of cosmic disorientation consequent on the demise of the Christian deity as a moral and metaphysical authority. Let's note two essential points about the madman's announcement.

The first is that the event is not about literal deicide. Nietzsche does not believe there ever was such a God, and even if there were It is by definition not the sort of thing that humans could kill. The death has rather to do with removing an all-encompassing source of meaning. It's easy to see how this works in the moral and political spheres. If God authors the definitive moral commands, then the commands die with the belief in God. If God legitimates the political order, then the monarch loses the justification for his or her rule when God is thought to expire. With the rise of secular moral doctrines and the separation of Church and State both things happened in the modern period. The *concept* of God now explains nothing in these spheres.

The same point applies to the workings of the physical universe. As late as the 18th-century, philosophers were still debating the role played by the divinity in sustaining the universe from moment to moment. The 17th-century French philosopher Nicolas Malebranche, no fringe figure at the time, argued that there could be no movement in bodies at all unless God were responsible for it. Malebranche's key claim is that it takes as much power to keep a thing in existence at any moment as it took to create it in the first place. But since the power of God is required to create the thing, Its power is also required to preserve it in place or move it.

Newton's three laws of motion remove the appeal of arguments like this. The first, the law of inertia—an object in motion will remain in motion unless an external force opposes it—makes continuous movement the *default* state of bodies. That is directly contrary to Malebranche's doctrine. Add to this the rise and development of heliocentrism and, a little later, of evolutionary biology, and you can appreciate how the old God might be getting a bit nervous about the precarity of Its position. In a mere handful of decades Malebranche's God goes from fussy workaholic to being automated out of a job. When Napoleon asked the French materialist philosopher Pierre Simon, Marquis de Laplace how God fits into his metaphysical system, the philosopher replied, "Sir, I have no need of that hypothesis." This sums up the first point nicely.

The second point is that just after the passage cited the madman goes on to notice, with some dismay, that the people hearing his message appear unready to absorb it fully. Indeed, we are informed that the crowd is composed of many who already do not believe in God, and this makes them dismissive of the madman's words. What is going on here? Is the madman only telling people what they already know? Not exactly.

Elsewhere, Nietzsche says that although God is dead Its shadow still haunts the world. Here's why. The great modern moral systems that arise in the Enlightenment period take over a key aspect of Christian morality while congratulating themselves on what they take to be their thoroughly secular foundations. That aspect is *impartialism*, the idea that we should take our interests to count as much as, but no more than, the interests of anyone who might be affected by how we decide to behave. Our moral gaze must, logically, extend beyond our partial concerns.

Perhaps you think this just sounds reasonable, and so it may be. But in his book *The Genealogy of Morality* Nietzsche demonstrates that it has its origins in the revolt of Christian slaves against their Roman masters. The Christians *inverted* the values of their overlords, in the process transforming meekness, humility, pity, weakness, selflessness and impartiality from vices to virtues. The master's values—selfishness, cruelty, partiality, pride, bodily power and so on—undergo exactly the opposite transvaluation, converted from virtues to vices. So Christian values emerge from a historically specific political contest and modern values bear traces of this origin. This opens up a blind spot in the system of modern values: its devotees tend not to grasp the ways their value judgments are also tools for projecting power. *Their* power.

The folks confronted by the madman are moral neophytes. For instance, few of them appreciate the contradiction between their impartialism and the slavery underpinning the global economic system. The inability to perceive such contradictions and tensions is the hallmark of moral immaturity. Has Nietzsche has committed the *genetic fallacy* here? The fallacy arises when we argue that some present phenomenon is debunked, or even completely explained, by pointing to its unsalutary historical origins. I don't think Nietzsche is guilty of the fallacy in this case. I'll say why in a moment, but first let me explain what is fallacious about this form of reasoning with an example from evolutionary psychology.

Sexual desire is without doubt a product of natural selection. We want our genes passed down the generations and the urge to procreate helps us achieve this, which is exactly why it was selected for in our ancestors. It's also plausible to say that the feeling of love has its origins in the primal push for genetic self-preservation. But things get much more problematic when we say that love can be explained fully by the subconscious urge to pass down our genetic material. For one thing, the 'explanation' can say nothing very illuminating about gay sex or love. For another, it cannot plausibly explain the love we feel for our children or grandchildren. But isn't this simply a matter of favoring those who will inherit some of our own genes? That's the official story from the evolutionary psychologist, but it's not particularly plausible.

For instance, in his popular account of evolutionary psychology, *The Moral Animal*, Robert Wright (1995, 174–176) informs us that we feel deeper grief if our child dies in adolescence rather than, say, at age 5. That doesn't sound crazy. After all, we have many more memories with the older child, so there is more to regret at her passing. That explains the difference, right? Wrong. The *real* reason, according to Wright, is that the older child is closer to procreation at that point, and so if she dies we feel the loss of opportunity to spread *our* genes more acutely than we do in the case of the 5-year-old's death. With the older kid, we were so close!

Any parent who thought it was preferable to lose a younger child rather than an older one *for this reason* would likely be a lousy parent, to say the kindest thing about them. Arguments like Wright's confuse historical (in this case, evolutionary) explanations with explanations of meaning. This is a fallacy because something like love has acquired its manifold meanings for us through the complex prism of culture, and in its meaning-giving capacity culture swamps biology. The genetic account of our emotions tells us something about our biological makeup while failing spectacularly to enlighten us about the emotions themselves.

In any case, Nietzsche does not make Wright's mistake, he does not commit the genetic fallacy with the death of God thesis as put forward in the *Gay Science*. He's just doing history (what he calls 'genealogy'). Although there were aspects of Christian morality Nietzsche did not like, the madman's announcement has nothing to do with critiquing the *content* of contemporary, secular value systems. He is not interested in either debunking or explaining away those systems with a story about their origins. Rather, Nietzsche is pointing to modern humanity's lack of self-awareness, in particular its failure to notice that its own system of values might contain tensions and contradictions because it was born, as all value systems are, in a power struggle.

In the last section of this chapter I will apply this lesson to us. For now, I want to pause for some reflection on the palpable weirdness of our situation. This may sound like an odd theme for a philosopher to explore, but it is central to the concept of bewilderment I'm trying to elucidate. Think again of Pierre, an exemplary character for us precisely because he is out of sync with his times. Or better, he is simultaneously in his time and out of it, occupying a strange middle place between the familiar and the unfamiliar, both at home and homeless. I'm interested in why he feels this way. As it happens, there's a word for his sort of mood: the uncanny.

An uncanny guest

One of the 20th-century's deepest interpreters of Nietzsche's work was the German philosopher (and National Socialist) Martin Heidegger. Heidegger's writing is notoriously difficult to decipher. His major work, *Being and Time*, is one of the great works of the century, but its German is so novel that translators have over the years struggled to render its meaning accurately. His later writing is, at times, even more difficult. However, scattered throughout his corpus are shorter and more accessible pieces, some of which are based on lectures he gave. His work on Nietzsche falls into this category. Heidegger saw Nietzsche as a kind of culminating figure in the history of Western philosophy, and he thought of himself as the thinker best able to interpret Nietzsche's epochal insights for a new technology-crazed century.

This task is summarized in Heidegger's interpretation of the madman passage. The passage, Heidegger informs us, heralds the advent of nihilism, the loss of the ground of all meaning. This event goes well beyond the demise of a single religious cosmology. The madman tells us that we have killed God, that we are Its murderers. And we know that the instrument of Its death is our science. Modern science desacralizes the world, sucks it dry of the value and purpose it had contained to that point. We can—indeed must—still create value systems but the death of God deprives them of the guaranteed stability they once enjoyed.

This is why the passage contains that dizzying string of rhetorical questions having to do with the loss of orientation. In early Christian cosmology, God was sometimes likened to the sun, the source of all heat and light. Stanford Professor of Italian Literature Robert Pogue Harrison (1993, 193) has suggested that our furious efforts over the ages to deforest the world were partly aimed at clearing these dark and tangled spaces in order to allow our sky God, embodied in the sun, into them (and also to provide us visual access to It). But of course the madman's message is that we are now moving away not just from this particular God, but "from all suns."

Science is able to perform this extraordinary act of robbing the cosmos of meaning because it declares that there is, and can be, no *purpose* in the cosmos. All religiously inspired cosmologies suppose exactly the opposite. They suppose that the divinity has infused the whole with meaning by, for example, setting it on some kind of inexorable historical path. Science, by contrast, finds nothing in that vast space but bodies pushing one another around, all this furious activity answering only to the immutable laws of nature. Science understands itself in almost entirely positive terms. It is that set of techniques and laws that illuminates the basic structure of the physical world. But from a Nietzschean perspective, the defining characteristic of scientific modernity is what this source of illumination removes from the world, the way it extinguishes the light of *meaning*.

This is nihilism, and for Heidegger grasping it fully requires noticing that it is the hardest thing to hear. Hence the madman's frosty reception or, for that matter, the tendency to label him 'mad' in the first place. However, Heidegger goes a step further. He argues that nihilism, taken this way, is an "uncanny guest" (1977, 63). What is the uncanny? Superficially, it is a slightly fusty linguistic stand-in for the creepy or weird. But these terms miss the full resonance of the concept.

The German term for the uncanny is *Unheimlichkeit*, literally 'un-homeliness.' If we take the ordinary understanding of the word—creepiness—together with the notion of un-homeliness, we draw closer to its essence. For the uncanny is the intersection of the familiar and the unfamiliar, the homely and the un-homely. It is the idea that although we cannot help but search for firm footing in the world, the places we occupy are always precarious. As I have pointed out, humanoid robots—those that are sort of like us but also, in their movements and speech patterns, still disquietingly other—are a good example of the kinds of objects that can evoke this feeling in us.

Here's another example, from the 1990 film *Exorcist III*. The film follows a police Lieutenant named William F. Kinderman, who is investigating a series of brutal murders that look as though they involve satanic worship. In one scene, Kinderman is looking for a witness in an insane asylum. He paces the communal area slowly, surrounded by the incoherent mutterings of the inmates gathered there. After a minute or two of this, the camera pans out, giving us a view of the whole room, including the ceiling. Then something scurries across the ceiling, though we cannot yet say what it is. The camera tilts upward to reveal a grand-motherly woman, complete with little gray hair-bun and floral nightie, crawling like a rat directly *over* Kinderman's head. She rotates her neck slowly away from the ceiling, smiles broadly at him and scurries away. Kinderman never sees her, and eventually exits the room.

I was just a kid when the first *Exorcist* movie came out in 1973. At the time I was an ambivalently pious and mostly incompetent altar boy at All Saints Catho-lic Parish just down the road from my house. I don't recall how I managed to see the movie without my parents finding out, but I'll never forget the nightmares I had for weeks after. More than 40 years on, I still can't quite shake the image of the satanically possessed Regan, who likes to pick on the apprentice-exorcist at her bedside, an emotionally frail young priest. She says the foulest things to him, often in the devil's own snarl.

None of the girls I knew from church said things like *that*, at least not in that tone of voice. I don't have nightmares anymore—at least not about the devil—but after *Exorcist III*, I never looked at my grandmother quite the same way. As with humanoid robots, and indeed Regan, the creepiness of the metamorphosed granny consists in the meeting of two elements. The scene is brilliantly con-ceived because the writers pick out the homeliest, the most familiar and the most comforting image they can conjure up, and stick a grinning demon in it. Simpler representations of evil—Dr. Hannibal Lecter, The Joker, Milton's Satan, you name it—cannot compete with this *uncanny* representation of it.

These illustrations of the uncanny are apt to make us believe that it is always a thing to flee. The more scientific term for the uncanny is cognitive dissonance, the feeling of inconsistency or contradiction among our beliefs, attitudes and behaviors. Because inconsistency and contradiction are uncomfortable psychic states we find various ways of not noticing them. The sociologist Kari Marie Norgaard (2011) recently wrote a brilliant ethnography of a small Norwegian ski village hit hard by rising temperatures, whose citizens exhibit this mindset.

Norway is a major producer and exporter of oil and is consequently very prosperous. But unlike, say, Saudi Arabians, Norwegians are also generally very environmentally progressive people. Norgaard describes the everyday ways Norwegians work through these dissonant elements of the national psyche, most of the time through one or another form of psychic flight. It's a subtle and culturally variable phenomenon, this deep sense of dissonance. But the conditions for it are now ubiquitous. Here are just two hair-pulling examples.

Along with over a thousand political jurisdictions across the globe, in May, 2019, the UK Parliament unanimously passed a motion declaring a climate change emergency. Good. At roughly the same time, however, a panel of British judges threw out a legal challenge to halt construction of a third runway at Heathrow airport, a project that will make it nearly impossible for the UK to achieve its Paris emissions reduction targets. A spokesperson for Friends of the Earth, one of the plaintiffs in the case, said the judges' decision is "outdated in an ecological and climate emergency" (Laville, 2019).

And now that decision has itself been declared illegal by the court of appeal. Industry players are poised to continue fighting for the runway, with the full support of the government. Indeed, the British transport secretary responded to the ruling by saying that "airport expansion is core to boosting global connectivity and levelling up across the UK." He added that the government also takes environmental concerns very seriously (Carrington, 2020).

The second example comes from the horrifying 2019–2020 bushfires in New South Wales, Australia. This disaster, demonstrably fueled by climate change, has devastated the non-human biosphere, destroyed hundreds of indigenous cultural sites and has so far cost the Australian economy approximately $100 billion. Australian novelist Richard Flanagan (2020) describes the situation this way:

> The name of the future is Australia. These words come from it, and they may be your tomorrow: P2 masks, evacuation orders, climate refugees, ocher skies, warning sirens, ember storms, blood suns, fear, air purifiers, and communities reduced to third-world camps.

So thorough has the damage been that some have argued we need a new word for it: omnicide, the death of everything. And yet, at the time of the fires Australia was being run by a government packed with politicians in the pockets of the fossil fuel industry. Spurred on by a denialist media machine owned in large part by the truly ghoulish Rupert Murdoch, there was a vigorous campaign to blame the whole thing on a few arsonists. Meanwhile, Australia is in the process of building 53 new coal mines and has almost no chance of meeting its very low Paris targets. With all of this in mind, Flanagan asks, "How can a country adapt to its own murder?"

I've read all the psychological literature on climate change, but when confronted with the contradictions exposed by examples like these, my eyes pop and I reach for a stiff cocktail. Useless reactions, I know. I feel like poor Pierre at one of those fancy balls, a bottle of Pol Roger at his elbow, his ears buzzing with the chatter of the experts holding forth knowingly on the significance of Napoleon for Russia's future. Even more pointedly, I feel like Nietzsche's madman and wonder how it is that we can appreciate neither the omnicide we are in the midst of perpetrating nor the tangled web of deceptions and evasions in which this act has entangled us.

I am bewildered, and my bewilderment is uncanny: I feel trapped in this existentially prickly space between a world that is quickly becoming hostile to our purposes and my *expectation* that it should instead be a place of comfort and solace. Having gone on long enough, at least for now, about the first half of this dichotomy we need to look more carefully at the second.

The lure of home

As we have seen, for Heidegger the madman announces an uncanny event, and in doing so to just this crowd something important is revealed to us about the human essence in the modern age. It is this: we are builders of homes. Not just literal ones of bricks and straw and wood, important as these are. I mean metaphysical homes, the sorts of homes that express or embody our values. For what are values after all? Our values are expressions of what matters to us. They embody our conception of what is *good*. As such they have a central, organizing role to play in our lives. They undergird our practices and institutions, acting as heuristics that enable us to make choices among the welter of options that is theoretically available to us at any moment.

We like to think that our choices, the important ones anyway, are justified and this means that we could in principle give reasons for them to anyone who asked (or to ourselves). Our values also speak volumes about how we believe we fit into a larger whole. For instance, if we consent to occupy a particular social role—serf, student, religious pilgrim, football player, CEO—we have implicitly endorsed a whole social order as well as the place we occupy in that order. We don't always feel good about the places we occupy, of course. This, however, proves the point. When we have the feeling that we should not be in this or that social role we express our disgruntlement by saying that we feel out of place here, that we don't belong in this role. The underlying assumption of statements like this is that we *should* feel this sense of belonging.

Now, it is crucial to notice that for most of our history, this order extended beyond our more or less circumscribed social world. It was cosmic, the immediate social world nested inside larger orders extending, as it were, all the way out. The Canadian philosopher Charles Taylor calls this our "social imaginary":

> By social imaginary I mean something much broader and deeper than the intellectual schemes people may entertain when they think about social reality in a disengaged mode. I am thinking, rather, of the ways people imagine their social existence, how they fit together with others, how things go on between them and their fellows, the expectations that are

normally met, and the deeper normative notions and images that underlie these expectations.

(Taylor, 2004, 23)

More abstractly, the social imaginary is comprised of three elements. The first is the norms that govern our everyday interactions, structuring our expectations. This is the way things simply are among a relatively cohesive group. But since things can fail to go the way we think they should at this level, we also sometimes change them, or want to do so, in accordance with an ideal. This is the second element. Finally, "beyond the ideal, there stands some notion of a moral or metaphysical order in the context of which the norms and ideals make sense" (Taylor, 2004, 25).

The tendency to think in terms of such structures of meaning and expectation is universal in our species, notwithstanding the profound differences among us—across both time and space—with respect to their content. We think this way for the very same reasons we build ordinary houses. We want a place in which to feel at ease, especially since so much of our lives seems to involve grappling with a world that is hostile to our purposes in one way or another.

We spend all day struggling against the unfamiliar, that which resists our efforts to shape or appropriate it, and want to come home to a cozy chair, a good meal, a warm fire, familiar faces and activities. Now transpose this set of images and expectations to the cosmic sphere. There's so much suffering in life, so much kowtowing, officially sanctioned bullshit and sheer drudgery. It can help us to bear all of this if our value system shows us that from the standpoint of the cosmic whole it *matters* that we get up every day and do it all over again, that it must be this way and is good that it is so. But the death of God complicates this enormously.

Again, the madman is not disappointed with those in the marketplace for having values and believing in them. He only wants them to notice that the foundation for those values—Taylor's third element, the "moral or metaphysical order"—has been removed, and that they should recognize this uncanny fact about their situation. The madman is informing them of a fundamental precarity in their world that they cannot see.

Too many people misconstrue Nietzsche on this issue. Throughout his corpus, Nietzsche is very clear about the importance of value-saturated enclosures to any flourishing life. The madman's hearers are not misguided, just objectionably self-satisfied. So we should not be too hard on these people, nor on the noble men and women in Pierre's circle. Without quite grasping it, they all occupy that uncanny intersection of meaning and unmeaning we have been trying to elucidate here. They are moderns. Of course they can't face the abyss that has opened beneath them, but how many of us can?

In any case, an additional source of precarity has been added to the one provided by modernity. Not only have we inherited the madman's message about the death of God, but the very air around us is becoming noxious. There's an invisible agent—a trace gas, if you can believe it!—that has been accumulating in an impossibly far-away place, the atmosphere, from there wreaking slow havoc on the basic workings of the Earth system. It's a bit like termites nibbling away at the wooden foundations of your comfy cottage, undermining its integrity one tiny bite at a time.

Folks in the 19th-century *could have* fully absorb what Nietzsche was telling them, while being entirely confident that nothing as sinister as this was going on behind their backs. But we no longer have this solace. I'm going to spend much of the next two chapters talking about this added source of precarity but we can get a glimpse of what's to come by underlining the subtle, but profound, change in how we think about this gas over the last few years.

Progressive governments, those that understand the urgent need to price carbon out of the energy market, have declared CO_2 to be a 'pollutant.' Think about the alteration of perspective involved in this otherwise insignificant linguistic shift. This gas, once thought of as not only innocent but life-giving, is now the enemy. We think of it, quite rightly, as the sinister byproduct of a global economic machine that is bringing the climate to its knees. For the whole 12,000-year history of the Holocene, we could afford to forget entirely about the work of CO_2. It was a stealth atmospheric agent keeping our Goldilocks planet nice and cozy, insinuating itself luxuriously into the greening world every spring to renew the base of the food chain.

But the dose, we know, is always the poison: there's too much of this stuff now, and it has therefore rushed to the foreground of our attention. Problematizing CO_2 this way has, in other words, lit up key features of the social imaginary. In doing so it has brought to our attention many of the contradictions structuring the social and political order, those exposed so artfully by Norgaard for example. Think of the existing social rifts that are now *widening* because of climate change: between North and South, the 1% and everyone else, white people and people of color, humans and non-humans, present and future people, teenagers and boomers. The grandfather of climate scientists, the late Wally Broecker, once described the new climate as an "ornery beast." This beast might see to it that we are torn apart by these divisions.

This, then, is the second source of bewilderment, layered onto the first. It is all our own. No human generation has ever had to deal with anything like it. The temptation to not notice it, or to fail to describe it the way we should, is itself ubiquitous, and often overpowering. There are manifold and seductive forces of denial in our culture, and I'm not talking only about the fools who believe that the Intergovernmental Panel on Climate Change (IPCC) is an international socialist plot or that 97% of climate scientists have deliberately fudged their data so that they can cash in on all the grant money for towing the 'alarmist' line about climate change.

No, I'm also talking about the way you and I choose to forget about our new world every time we rise to the alarm, plug in the coffee maker, prepare our breakfast, get our kids to school, drive to work, push our pencils, have drinks with colleagues, come home, make dinner, watch some Netflix, go to bed and do it all again the next day. In other words, carry on as though something of monumental significance were not unfolding all around us.

By now we all *know* there's a connection between how we live and the increasingly weird weather. Even so, I'm suggesting, we need the spur of uncanny events to appreciate all of this appropriately. The mounting disasters of climate change are beginning to perform this awful, vital cultural function. We *are* slowly waking up, but I predict we are going to have a hard time processing the world we encounter as we shake off the cobwebs of our long Holocene slumber.

The attractive thing about Pierre is that he sometimes emerges from *his* periodic encounters with the deeper forces governing the world with a renewed sense of strength and responsibility, an invigorated feeling for the beauty of the whole. For example, after witnessing the summary executions of Russian soldiers in the French prison camp, Pierre is appropriately shattered but somehow his faith in life is renewed after a long evening's conversation with a fellow prisoner and simple peasant: "For a long time Pierre did not sleep, but lay with eyes open in the darkness . . . and he felt that the world that had been shattered was once more stirring in his soul with a new beauty" (Tolstoy, 1994, 1381). This experience is never purely subjective in him, a mere feeling. Rather, it provides him with a kind of moral clarity that affects the way he consorts with others. It's this possibility for moral *transformation* buried in our times that I want us to unlock, and whose various dimensions the rest of this book explores.

So here's the main message of this chapter: we need to learn how to accept our bewilderment. Though we cannot suppress it altogether, *right now* the need to feel at home is mostly likely a form of flight. We should be suspicious of it. All species of climate change denial—from the hard-core dismissal disseminated by Fox News to those who claim to care but do nothing to change their ways—are premature attempts to declare solid the foundations of our civilizational dwelling. But they are not solid, and we must expose the cracks ruthlessly. This is not to fetishize confusion. If someone asks you to torture a child for no reason, it would be perverse to claim to be confused by the choice facing you, as though you should be praised for your complexity in deciding you'd like to think this thing over. That's naked posturing.

But now the world just *is* confusing. The ugly compromises and bald inconsistencies of our late-Holocene social imaginary have been exposed harshly by the climate crisis, and this has produced a deep sense of anxiety and confusion among us. Let's try to abide in and understand this complex mood. The existential therapist Irvin Yalom has argued (1980) that it can be important to resist attempts to flee or deny anxiety. Instead, we should "plunge into the roots of . . . anxiety for a period of time, experiencing heightened anxiousness" (206). This is what we all need right now, to see our anxiety as a chance to clarify what really matters. It's easy to laugh at those, like Pierre and the madman, who appear dazed and directionless. I prefer to think of them as representations of a hard-won moral maturity.

Conclusion

I can already hear some saber rattling from those who will want to accuse me of dithering. Let me therefore close with two points. The first is that to recommend bewilderment—Yalom's "heightened anxiousness"—is not to counsel fatalistic withdrawal or passivity. There's a trend in our culture in this general direction, for instance by those, like the American novelist Jonathan Franzen (2019), who embrace what they see as the inevitability of climate apocalypse. Though I think it is eminently worth engaging, I don't want my argument to be mistaken for an endorsement of this view *if* it is taken as a counsel of political despair (which, to be fair, is not Franzen's aim). We can act resolutely while abiding in existential anxiety.

The second point reiterates something I've already said, namely that bewilderment is not an end in itself. It is merely an antidote to complacency. It exposes the social imaginary's contradictions, and thus opens a path to moral clarity. Notice that in the earlier quote, Yalom stresses that anxiety should be indulged only "for a period of time." Parts II and III of this book represent my attempt to move beyond bewilderment, though this mood will doubtless pop up perennially (as with COVID-19). In any case, the point here is straightforward: let's fight the climate crisis with due vigor *and* look for ways of snapping ourselves out of our current bewilderment, all the while being alert to the possibility that what looks like clarity might be yet another form of denial and evasion. That's a complicated enough challenge for now.

You might be inclined to press me on two further issues, both of which are connected to exposing what Yalom refers to as the "roots" of our anxiety. The first has to do with divine foundations. Nietzsche has not exactly disproved God's existence, only showed how modern science has rendered It redundant. The history of ideas from Anselm of Canterbury (1033–1109) to contemporary Intelligent Design Theory is replete with arguments purporting to prove the existence of God. The Nietzschean chastening notwithstanding, don't they, or some of them, establish the case for the divinity as a secure foundation? If not, is there perhaps a non-foundational way to believe in God?

Second, and speaking of foundations, why can't Nature provide them? Maybe the death of God is something of a boon because it finally allows us to measure our doings against what Nature intends for us. Time to pay a few explanatory debts by looking more carefully at the crucial issue of metaphysical foundations.

References

Carrington, D. (February 27, 2020). "Heathrow Third Runway Ruled Illegal Over Climate Change." *The Guardian*. Retrieved from: www.theguardian.com/environ ment/2020/feb/27/heathrow-third-runway-ruled-illegal-over-climate-change. Accessed March 6, 2020.

Cohen M.A., and Zenko, M. (2019). *Clear and Present Safety: The World Has Never Been Better and Why That Matters to Americans*. New Haven: Yale University Press.

Flanagan, R. (January 25, 2020). "How Does a Nation Adapt to Its Own Murder?" *New York Times.* Retrieved from: www.nytimes.com/2020/01/25/opinion/sunday/australia-fires-climate-change.html. Accessed January 31, 2020.

Franzen, J. (September 18, 2019). "What If We Stopped Pretending?" *The New Yorker.* Retrieved from: www.newyorker.com/culture/cultural-comment/what-if-we-stopped-pretending. Accessed January 27, 2020.

Harari, Y.N. (2016). *Sapiens: A Brief History of Humankind.* New York: Signal.

——— (2017). *Homo Deus: A Brief History of Tomorrow.* New York: Signal.

Harrison, R.P. (1993). *Forests: The Shadow of Civilization.* Chicago: University of Chicago Press.

Heidegger, M. (1977). "The Word of Nietzsche: God Is Dead." In *The Question Concerning Technology and Other Essays,* William Lovitt (ed.). New York: Harper Torchbooks, 53–114.

Laville, S. (May 1, 2019). "Friends of the Earth to Appeal Against Heathrow Judgement." *The Guardian.* Retrieved from: www.theguardian.com/environment/2019/may/01/high-court-dismisses-attempt-to-block-third-heathrow-runway. Accessed January 26, 2020.

Nietzsche, F. (1974). *The Gay Science,* translated by Walter Kaufmann. New York: Vintage.

Norgaard, K.M. (2011). *Living in Denial: Climate Change, Emotions and Everyday Life.* Cambridge, MA: The MIT Press.

Pinker, S. (2018). *Enlightenment Now: The Case for Reason, Science, Humanism and Progress.* New York: Viking.

Potter, A. (February, 2018). "Stephen Pinker, and Is Enlightenment Enough?" *Literary Review of Canada.* Retrieved from: https://reviewcanada.ca/magazine/2018/02/steven-pinker-and-is-enlightenment-enough/. Accessed April 14, 2019.

Shapiro, B. (2019). *The Right Side of History: How Reason and Moral Purpose Made the West Great.* New York: Broadside Books.

Sophocles (2007). "Antigone." In *The Moral Life,* Louis P. Pojman (ed.). Oxford: Oxford University Press.

Taylor, C. (2004). *Modern Social Imaginaries.* Durham: Duke University Press.

Tolstoy, L. (1994). *War and Peace,* translated by Constance Garnett. New York: Modern Library.

United Nations. (2019). *Nature's Dangerous Decline Unprecedented; Species Extinction Rates Accelerating.* Retrieved from: www.un.org/sustainabledevelopment/blog/2019/05/nature-decline-unprecedented-report/. Accessed May 7, 2019.

Wright, R. (1995). *The Moral Animal: Why We Are the Way We Are: The New Science of Evolutionary Psychology.* New York: Vintage.

Yalom, I. (1980). *Existential Psychotherapy.* New York: Basic Books.

2

FOUNDATIONS: GOD AND NATURE

In mid–April, 2019, I watched (on the Internet) the spire of Paris's Notre Dame cathedral collapse through the roof, engulfed in flames. The reaction of Parisians to this event was telling, many of them weeping openly in the streets. Of course, it was, and is, a beautiful building and so these displays might be understood as reactions to the sudden and violent destruction of something of surpassing beauty. But there's surely more to it than that. I think the fire was an assault on what many people in France (and elsewhere) still think of as a basic point of cultural orientation. Paris Point Zero is located in the square just outside the cathedral and was reckoned to be the heart of both the city and the country, all points radiating out measurably from it. The church's foundation stone was laid by Pope Alexander in 1163, thus solidifying and consecrating a spiritual bridge between the two end points of the Great Chain of Being, rock and God.

I'm fond of that old saying that our house is our castle and our keep. We no longer use 'keep' as a noun. I wish we did because it evokes a rich idea of dwelling, combining the notions of personal and communal security on the one hand and safe storage of our most precious belongings on the other. In the past it was used mainly to describe a castle's inner stronghold, its stony nucleus. It's a material foundation and place of comfortable return from life's dangerous byways. Notre Dame is a cultural keep in the complex sense I want to emphasize here: it connects a whole people to each other, to the surrounding topography, to the built world and to the divinity. It is a foundation, place of security and point of orientation. And because it is both material and spiritual, it also encompasses and protects a peoples' values, those that are professed at fixed times inside this baroquely imposing keep. How should we understand this expansive sense of foundations, one whose most important function is to ground our collective values and practices?

The first thing to say is that reliance on foundations, so construed, goes beyond trust. Trust is possible, or perhaps required, only where there is some degree of uncertainty. I trust the electrician I have hired to do the job we've agreed on, but I also recognize that he might, for one reason or another, fail in his task. The philosopher Annette Baier captures this aspect of trust nicely, defining it as "accepted vulnerability to another's possible but not expected ill-will (or lack of goodwill) towards" us (1994, 99). Trust is an ambiguous attitude because it involves reliance on something we cannot fully control, predict or understand. Like love, trust makes us hostages to fortune. It's a tacit admission that the future is open, perhaps frighteningly so.

By contrast, I don't trust that the next bachelor I meet will be unmarried, for it can be no other way. This is a matter of expectation, the recognition of a constraint on the future. Given what has been argued in the previous chapter, it probably won't surprise you to hear that I believe God can provide no such foundation for our values and practices, and this is because we cannot know with certainty that such a Being exists. However, there's a different way to think about God, a way that requires no foundations and is essentially a matter of trust. We'll spend some time making this key distinction. The upshot, we'll see, is that while God can indeed provide no secure foundations for our values and practices, this fact alone does not undermine all basis for belief in God.

That's the first half of the chapter. In the second half, we come to the second potential foundation for our values and institutions: Nature. This looks like it might be just the ticket for an age with no other foundational resources, but, as we will see, it has become just as problematic in this respect as God is. The main concern is that in our new epoch, the Anthropocene, traces of human artifice are everywhere in the Earth system. The boundary between the natural and the artificial, culture and the wild, is therefore no longer discernible. And if we can't make out that boundary, then it is hard to see what it could mean for our values to have their foundation in Nature. To do its job properly, a support must be distinct from that which it supports. But if it's misguided to look to God or Nature for foundations, then we're in a uniquely tricky metaphysical spot. We need to clarify what this means.

Science and religion

Back in the 1990s and early 2000s a sometimes acrimonious debate took place between the so-called new atheists and various defenders of the faith. These atheists—Richard Dawkins, Christopher Hitchens, Sam Harris and Daniel Dennett—were dubbed the four horsemen by their critics. In fact, they were only updating arguments that have been made numerous times since the Enlightenment. They made lots of claims, but the overriding point was to show us, once again, that modern science destroys the basis for belief in a transcendent deity.

The responses too were diverse, but tended unsurprisingly to converge on the antithesis of just this claim. Science and religion, so went the reply, are fully compatible. Call this the compatibility thesis. I think it is basically correct, but it is crucial to articulate it properly. It cannot rest on the notion that we can know with certainty that this Being exists. And if God's existence cannot be proven, then God cannot be a metaphysical foundation. To get our bearings on these complex issues, we begin with some basic epistemology (the theory of knowledge).

Philosophers make a distinction between truth-functional and non-truth functional sentences. 'It is raining outside' is a truth-functional sentence. Truth-functional sentences are either true or false, and in most cases we have established methods for figuring this out. If I want to prove to you my claim about the weather, I can just open the drapes and show you what's going on out there. The philosophical lingo for this is that there are mostly unproblematic 'truth conditions' for claims about the nature of this or that bit of the world. The exclamation 'Eeeaww!' emitted by me after smashing my thumb with a hammer is not truth-functional. It's just an expression of pain. Of course I could be faking the pain, but in this case it's not the *expression* that's false. It's the associated truth-functional claim (which I may or may not actually articulate), 'I am feeling pain in my thumb right now.'

The theist's claim that God exists is clearly truth-functional and has historically almost always been thought of this way. Over the ages, innumerable attempts to prove the claim have been made. Most fall into one of two categories, ontological and cosmological. Ontological proofs attempt to show that God's existence follows from Its essence. God is definitionally perfect, and since existence is a perfection, God must exist. Cosmological proofs begin with some observation about the world, then work back to the existence of a perfect Being as the only, or best, explanation of that thing. For example, we observe that nature is full of design, that all design presupposes a designer, but that the complexity and beauty of natural design—think of the human eye, the go-to example of contemporary Intelligent Design theorists—far outstrip the human capacity for design. Therefore a divine designer must exist.

There are endless permutations of each of these two basic argumentative types. If I were writing a different book, I'd refute them all, one by one. But, first, that would take far too long, and I don't have the space here for digression on that scale; second, it's already been done anyway, and I loathe repetition. We did not really need the arguments of Dawkins *et al.*, much as I enjoyed reading those books. This is because the great 18th-century Scottish philosopher David Hume (1711–1786) conclusively refuted the main arguments for God's existence in his *Natural History of Religion* and *Dialogues Concerning Natural Religion*. That the same theistic arguments continue to emerge in the long wake of Hume is a sad testament to human obstinacy. So I will come at the issue in a more roundabout way by asking the following question about all of these alleged proofs. If they actually worked, definitively establishing the existence of God, shouldn't we all be theists by now? The answer is yes, and here's why.

Humans are not only *not* incorrigibly irrational, most of us love a good proof when we encounter it. We are cognitive hedonists. Assenting to an indubitable, or even just highly plausible, argument or proof *feels* really good. Mathematicians get this, but the point applies beyond that uniquely elegant discipline. I've been teaching philosophy for about 20 years, and I still get a chill when I present my students with a clear argument on some topic and suddenly see the light of understanding go on in them. I recognize the signs: the brightening eyes, the slowly spreading smile, the twitch in the knee, the rush to write it out or tap it in before the insight vanishes.

But the notion that this is all about a good psychic feeling doesn't capture the essence of the experience. There's something deeper going on, something connected to the complex relationship between compulsion and freedom. Come back to the example of mathematics. René Descartes (1596–1649), whose thinking we will explore in depth in Chapter 6, was one of 17th-century Europe's finest mathematicians. One of the things that impressed him most about math is the way we are *forced* to assent to true mathematical propositions. Not forced externally, by someone telling you that you had better assent or else. No, this is a kind of internal compulsion, one involving the relation between the contents of the mathematical proposition and the way the mind works in the mode of assent or denial. When confronted by a mathematical proposition like '2+2=4' we spontaneously affirm its truth, assuming we understand its constituent terms. Indeed, for propositions like this there seems to be little difference between comprehending what the proposition means and assenting to it. There's almost no light between those two mental operations.

An even clearer example of this phenomenon is Descartes' famous proposition, 'I think therefore I am.' The idea behind it is disarmingly simple. Even if there's an all-powerful demon bent on deceiving me about the things I take to be most certain, this being cannot make it the case that I don't exist if I think that I do. The proof is that the very act of trying to affirm the truth of this hypothesis—'I don't exist'—refutes it. For how could the deceiver deceive me if I weren't around to be deceived? Again, we are, it seems, *compelled* to assent to the truth of the proposition 'I think therefore I am.'

This fact forces us to recognize a fundamental paradox: that we may be most free when we are most compelled. It's a paradox because we tend to think of freedom as the absence of compulsion, its ultimate antithesis. This makes sense with external compulsion—think of a command backed up only by a threat— but not with the right kind of internal compulsion. In the cases of internal compulsion we are considering here—assenting spontaneously to clear proofs—we seem to be doing what we really want to be doing. But isn't this what acting freely means? It seems irrelevant that we could not have done otherwise.

What does any of this have to do with God? The theist might say that in refusing to accept this or that proof of God's existence, the rest of us are displaying either intellectual stubbornness or incorrigible stupidity. I'll let the suggestion about our stupidity pass since it's only an unwarranted insult. Are we being stubborn, then, boldly denying that which we would apprehend clearly but for our intellectual pride? The problem with this claim is that good proofs just cannot be resisted on such flimsy grounds.

In *Notes From Underground*, Russian novelist Fyodor Dostoevsky (1821–1881) portrays a character who tries to do just this but his efforts prove to be futile and pathetic. In deciding to oppose clear mathematical truths and the like, it becomes impossible for him to make genuine sense of reality. Because he cannot sustain this stance sincerely, he retreats further and further into a solipsistic world—symbolized by the dingy basement flat he occupies—that is indistinguishable from madness. That's what the attempt to willfully deny clear truth brings.

The average atheist is nothing like this sorry fictional figure, flailing at the walls of allegedly certain truths. As far as psychological types are concerned, the underground man is the exception who proves the rule. I conclude that if the putative demonstrations of God's existence were any good, humanity would long ago have discovered this and we would be well on our way to the promised land by now. That we are nowhere near doing this is, to my mind, solid evidence that the arguments are poor.

However, this does not usher the deity off the philosophical stage. It shows only that this Being's existence cannot be rationally demonstrated. I've been suggesting that there's a tight connection between God conceived of as a foundation and our ability to prove Its existence. Having dismissed the idea of proof in this case, we are evidently deprived of the divinity-as-foundation as well. So if belief in God is to be justified at all, God must be conceived of as *absent*, in the sense of being inaccessible to reason. Think of this in terms of the distinction between expectation and trust, from this chapter's Introduction. You can't sincerely entertain *expectations* about or towards something that might not be there, though it is still open to you to *trust* in this possibly existing Being in one way or another.

This, then, can serve as the basis for a non-foundational belief in God. How can we conceptualize a vanished God? The notion that God is essentially hidden has deep historical roots. This is the *Deus absconditas*, who first appears in the book of Isaiah (45:15): "Indeed you are a hidden God, you God of Israel, the Savior." Much later, Luther made a big deal of the idea in his theology. It's a strange concept. Why believe in something that goes out of Its way to avoid detection? What's with this divine coyness?

The original articulation of the *Deus absconditas* concept was not about whether or not we needed God to help us do physics. It was, rather, meant to make believers reflect on the reasons for trusting a God who *seems* to be indifferent to human suffering (since It does not prevent such suffering, but presumably could). Apropos of this theme, the great American writer Flannery O'Connor said this:

> [People] think [religion] is a big electric blanket, when of course it is a cross. . . . You arrive at enough certainty to be able to make your way, but it is making it in darkness. Don't expect faith to clear things up for you. It is trust, not certainty.

> *(1979, 354)*

The images here are telling. Belief in God is, explicitly, not a matter of perfect certainty and the comfort it provides. We shouldn't seek proofs for God's existence. Belief in this Being is instead a voluntarily assumed burden, taken on in uncertain and vulnerable trust. Even if certainty has vanished, religious people can still trust in their God to show them a way through the world. But because it leaves the religious believer exposed to the whims of a God whose will cannot be known, such trust is constitutively anxiety-ridden. All the great Christian existentialists from Søren Kierkegaard (1813–1855) to Gabriel Marcel (1889–1973) have understood this.

This does away with, or at least bypasses, the attempt to understand belief in God as a matter of formulating the right truth-functional sentence. As we have just seen, that approach always demands a proof. If I say it's raining outside and you are skeptical about my claim, it's up to me to prove to you that the claim is right. But I'd hardly bother doing this, and you'd likely not bother asking me to, if we both knew that the current state of the weather could never be demonstrated. The *Deus absconditas* is like this.

I'm not advocating for belief in this God (for what it's worth, I don't believe in It or any other version of It). What these points highlight is rather that *reflective* atheists and theists are not metaphysical poles apart. They share the knowledge of fundamental precarity and both have the scars of anxiety to prove it. This is potentially fertile common ground between them, valuable real estate in the context of the climate crisis. While figures like Canadian physicist and evangelical Christian Katharine Hayhoe and Pope Francis (among many others) are overtly trying to get Christians to see that they have a moral duty to address the crisis, they are also doing yeoman's work building bridges between theists and atheists. The language of common values across cultural divides comes up often in Hayhoe's work, for example.

But we mustn't move too quickly. Trusting in God is justified only if it is not actually *irrational* to believe in this Being, hidden or not. The theist cannot justify believing in something that we can prove does *not* exist. As it happens, however, there's nothing for her to worry about on this score, for God's non-existence is no more susceptible to proof than is Its existence.

Of wagers and water buffalo

The great French philosopher and mathematician Blaise Pascal (1623–1662) asks us to imagine a coin being flipped at the end of the universe: heads God exists, tails It doesn't. The problem, he averred, is that reason cannot decide the issue for you. It's a wager. Pascal goes on to show that it's good to bet that God exists since if you bet that It doesn't when It does you will roast in hell for all eternity, whereas if you bet that It does and It doesn't you will at worst have foresworn a life of sex, drugs and rock and roll, things which aren't that important anyway.

This is true *even if* the coin is heavily biased towards tails, since in this case we are still talking about betting one life—this one—against an infinity of possible lives. As Pascal puts it, "the finite is annihilated in the presence of the infinite." So long as the probability of God's existence is not zero, it is rational to bet that It exists. "Reason," Pascal says, "can determine nothing here," because there is "an infinite chaos" separating us from what's out there. One way or the other, a leap is required (Pascal, 2008, section III).

Pascal was a Jansenist, a Catholic sect that learned a lot from the Protestants about the evils of sensuality. There's a truly terrible 2004 movie called *The Libertine*, set at about this time. In it, Johnny Depp plays John Wilmot, 2nd Earl of Rochester, a wine and sex besotted raconteur in the English court of Charles II. Pascal's argument is directed at this kind of guy. In a way it's quite ingenious, because these are not hyper-rational atheists, people demanding a good argument to turn their heads.

For starters, they generally like to gamble, so the crafty religionist can, as it were, meet them at the table. Because they are already so busy in the wee hours, they are indifferent to most of the questions that keep the theologians up at night: how many angels can dance on the head of a pin, whether the eucharist is really the body of Christ or only metaphorically so, how that whole trinity thing is supposed to work. Blah, blah, blah.

Except, of course, for the nagging issue of eternal pain. As a rule, the more you like pleasure, the less you like pain. If you're Wilmot's kind of sensualist you'll therefore want to prick up your ears when the priest mentions the possibility that the path you are on will result in eternal pain for you. So long as the issue can be contemplated over a brimming flagon of claret, it's clearly worth your time. "Christ almighty, I'm having fun," Wilmot might say under such discursive pressure, "but I'm not one of those masochistic fellows. If this life's going to result in everlasting torture, I'm out!" Time to head for the confessional. Clearly, Pascal understood his audience.

This appeal is too culturally parochial and historically specific to move most of us, but the bit about the coin toss at the end of the universe should. Although some of them fail to appreciate the point, no atheist is entitled to the claim that God does *not* exist. The reason is simple: you can't *prove* a negative existential claim about objects, whether gods or rocks. To see why this is the case, let's begin with a comparison. We *can* prove the non-existence of married bachelors. That's because of the definition of the relevant terms: the property 'unmarried' is contained within the concept of a 'bachelor.' Take it out and you just don't have a bachelor any more.

We get this result only because we are talking about a concept rather than an object. The problem with objects is that we cannot disprove the existence of any of them. This sounds counterintuitive so here's an admittedly outlandish example to clarify the point. I'm pretty sure there's no sparkly blue water buffalo sitting in the room I'm in right now. Can I prove this? Looking in the corner reveals no such beast, but what if there's something wrong with my eyes (not to mention my nose and ears)? What if, unbeknownst to me, I have a rare genetic disorder resulting in the specific inability to see anything sparkly blue or—horror of horrors—to perceive any member of the species *Bubalas bubalis*?

Consumed by contemplation of this awful possibility, I might rush into the corner of the room where I suspect the animal is hiding, hoping to *not* run into it, a non-event that would surely confirm its non-existence. But this is something I would probably not have bothered doing had I known that being sparkly blue confers on the water buffalo an ability to shrink in size when rushed at in this manner. This allows it to scurry away undetected whenever it wants, like when a hairy, half-crazed philosopher with a weird visual impairment is running towards it.

It's starting to look impossible for me to verify the non-existence of this animal in the room. Note that all that is required to set this chain of worries in motion is the bare possibility that they might be true. There is nothing *logically* suspect about suddenly shrinking water buffalo or genes that prevent the perception of some species but not others, even if physics and biology cannot make much sense of these happenings.

The potentially discouraging thought about the water buffalo can be generalized to any object in the universe, or indeed beyond it. In the case of the water buffalo, we are content with a probability claim. Given everything else we know about the way the world works, it is highly unlikely that there's one in the room right now. A good enough conclusion for all practical purposes. But there's an important difference between God and the sort of belief we have just been investigating. When it comes to God, Pascal seems to be right about the coin toss, which is not something we'd say about the water buffalo.

The difference between the two is that we expect claims about the existence of material objects to be vindicated through ordinary sensory means. If it's there, it can be sensed. If it can't be sensed, then there's a high probability—not a certainty, as we've just seen—that it's not. But if the entity in question is hidden *by its nature*, then this ordinary way of vindicating Its existence can't be very enlightening. In this case, it seems to me, we are faced with something like a coin toss. And if you choose to believe the thing exists, then O'Connor and Pascal are right: this is a matter of living your life in uncertain trust.

There's one more important thing to say about Pascal, namely that he has an especially arresting way of describing the main source of existential anxiety. He talks about the terror inspired in him by contemplation of what he calls the universe's "infinite spaces," those revealed by the relatively new technologies of telescopy. These spaces, he supposes, are a literal void, incapable of supporting anything. Beyond them might be a void too. But Pascal nevertheless resolves to exist in doubt-and-anxiety-ridden vulnerability, with no epistemic access to foundations.

If we understand this stance as the determination to go out and build places to dwell, structures that both express and enclose our deepest values, then we should also say that Pascal clearly gets the uncanny. He understands that life is all about living and creating resolutely in the shadowy meeting place between the familiar and the unfamiliar.

In these two sections, I've been urging a way of thinking about the compatibility thesis that should be entirely amenable to the atheist. Science does not, and cannot, disprove the existence of a transcendent deity. At most it can push this Being out of a job running the world. And it has done this. But what it has left entirely undisturbed is that coin of unknowable bias being flipped, silently, at the end of the universe. Boosters of science therefore cannot blithely declare that the person who decides to pick heads—God exists—is being irrational. That's what dogmatic atheists like the four horsemen and their legions of supporters some-times fail to appreciate. It renders many of their broadsides tiresomely tin-eared.

Everybody, theists and atheists, can live with integrity without thinking of the deity as a secure metaphysical foundation. In principle, we can all draw together and face as one the fundamental precarity of existence. It is more important to say this now than it ever has been for us. As I suggested earlier, if we don't come together in very short order then we are probably going in the opposite direc-tion: toward greater and greater fragmentation along those lousy lines traced out by race, creed and class. It may be that looming climate disaster will force an overdue rapprochement between atheists and theists.

I certainly hope so. In Chapter 5, I'll return to these themes, this time in an effort to be more philosophically precise about the connection between religion and care for the biosphere. For now, it's time to look at Nature, the second poten-tial value-foundation.

Nonsense on stilts

The point of talking about authoritative or foundational sources of meaning is to highlight the sense in which the source picked out can be a model for us, for our actions, our artworks and our institutions. It provides us with an ideal on the basis of which we can confidently build the various items of our cultures, includ-ing our values. Nietzsche's madman seems to shatter the possibility of anything at all playing this sort of role.

But maybe there's a potential source of authority he overlooks. What about Nature? This seems promising. After all, we often criticize this or that behavior or cultural production by calling it unnatural. We might think this locution is too old-fashioned to be taken seriously, but people still talk this way. On the back of the novel I'm currently reading, the publisher informs me confidently that the story in these pages "holds a mirror up to Nature." Here, Nature is the model for art. That's a kind of foundation, and the appeal to it is not obviously incoherent, so in this section we'll take a closer look at this idea.

The attempt to ground our most basic values and institutions in the natural order goes back a long way. In a relatively recent version of it, the 18th-century political theorist and one of the Founding Fathers of the American Republic, Thomas Paine (1737–1809) declares that "all the great laws of society are laws of nature" (1961, 400–401). That was a common way of thinking in this period. We find it in the political philosophy of the English philosopher John Locke, for example, for whom the right to private property is a law of nature, there *before* the institutions of government are constructed and meant to constrain those institutions in specific ways (Locke, 1980). This view has been described aptly as "possessive individualism" (Macpherson, 1962).

Among political philosophers at this time, the discussion was largely confined to what *human* nature is like. But in the wake of Darwin's work in the 19th-century justifications for striving individualism were sought out in *all* of nature. The new claim was that the individualist behaviors of all living things—from trees competing with their neighbors for every scrap of sunlight poking through the canopy to humans sweating it out on the trading floor—can be understood in accordance with nature's unchanging ways. For some, like Herbert Spencer (1820–1903) and Friedrich Hayek (1899–1992), this natural fact becomes a justification for unfettered markets and the denial that the job of governments is to aid those deemed less fit.

Much further back, Aristotle does much the same thing with slavery. He believes that this institution, widespread in the ancient Greek city-states, is justified because some people have slavish natures. And what a happy coincidence for the masters that those with putatively slavish natures just happen to be the enslaved ones! Clearly, these folks are not fit for any other social or economic function. Look at them, behaving just like slaves. I do hope you see where I'm going with this openly sneering description of the view under consideration. But sneering only gets you so far, so let's look at the arguments.

Suppose you think Spencer and company are basically right, Aristotle obviously wrong and that the general appeal to Nature as a foundation for judgments like these is legitimate. In this case, you would be saying that Nature (a) grounds an aggressive individualism, but (b) does not ground slavery. In other words, it justifies institutions based on (a), but not those based on (b). What would your evidence be for each half of that claim?

Take (b) first, the bit about slavery's alleged *un*naturalness. The problem here is that we don't have to look too far to find examples of slavery in nature. For example, there are at least six ant species—like *Polyergus breviceps*, the Amazon ant—that enslave other ants. There are various methods for accomplishing this. One is that a queen and her troops will invade another nest, usually in the summer when there are plenty of pupae in it, kill all the adult ants there, lay her own eggs and wait for the pupae to emerge. The wee ones will assume the foreign queen is their real queen and will then be employed for their whole lives raising her brood and helping conduct slave-seeking raids on other nests. All with no chance of reproduction for themselves. Similar behaviors can be found among certain species of bees, beetles and crickets (Sekar, 2015).

What about (a), which has become the ideological basis of unconstrained capitalism? I'm still amazed at how often I hear a version of this from my students, usually when I ask them to write me an essay on Karl Marx. The line, in brief, is something like this: Marx has some smart things to say about how capitalism functions but he was too idealistic about the possibility for human cooperation because we are all *by nature* individualistic and competitive. Capitalism is justified because, just like my novel, it holds a mirror up to nature.

Unfortunately, we can do the very same thing with these claims that we did in the case of slavery. Peter Wollheben (2016) provides an exhilarating glimpse of some of the latest discoveries in the science of forestry. His key insight is that forests are complex communities of organisms engaged in various form of communication and symbiosis. The term ecologists use for this is 'mutualism,' and it is the defining organizational characteristic of mature ecosystems.

In a mature forest, the tangle of *natural* cooperation among plants, fungi, insects and animals is in stark contrast to what goes on in planted forests:

> Because [the roots of planted trees] are irreparably damaged when they are planted, they seem almost incapable of networking with one another. As a rule, trees in planted forests like these behave like loners and suffer from their isolation. Most of them never have the opportunity to grow old anyway. Depending on the species, these trees are considered ready to harvest when they are about a hundred years old.
>
> *(48)*

How curious. The trees *we* put into nature, for purely economic reasons, are mangy individualist loners, whereas naturally mature forest systems are full of eager-to-network mutualists! All of this rather turns possessive individualism, especially the variety rooted in evolutionary theory, on its head, doesn't it? The real moral of the two examples—the ants and the trees—is that Nature provides instances of symbiosis *and* competition, slavery *and* equality. Any claim that it supplies us with a single unambiguous model for our values, social relations or institutions—including our economic system—is a patent exercise in cherry-picking.

Come back for a moment to those student papers on Marx. The views they express about what Nature is like are exactly what Marx himself said would be the dominant view *in a capitalist society*. This point is connected to my dismissal of Aristotle. What people invariably do in these cases is take the norms and values dominant in their social order and project them back onto the natural world. They do this because they believe that the comparison justifies or legitimates their social order. And it's because they feel so *comfortable* in the world they have inherited that they feel entitled to suppose that this must be what Nature intended. Because it is so ubiquitous, let's take a more extended look at this way of thinking.

In her novel, *Room*, made into a film in 2015, Emma Donoghue tells the story of a man, Old Nick, who keeps a woman as a sex slave in a tiny underground bunker constructed in his backyard. Over the course of her captivity, the woman—known only as Ma—bears a son she names Jack. Jack grows into a boy whose *whole world* is this room. When they eventually plot their escape, Jack is terrified by the prospect of making his way through the world beyond the room. Now, would anyone suppose that the *comfort* Jack feels in his underground world proves that children are by nature meant to live out their lives in cramped captivity? That this is natural for them? Of course not. So why make the appeal in other, related cases?

For example, in *The Subjection of Women*, the late 19th-century philosopher John Stuart Mill, in collaboration with his wife, Harriet Taylor, took on Victorian sexism in much the same way. The argument of the patriarchs was that proscriptions on women with respect to employment, education, voting, holding political office and so on were justified because nature made 'the fairer sex' unfit for these roles. In fact, so went the argument, apart from a few suffragette shit-disturbers, most women seem entirely content with their subordinate status. Surely that's proof that this is natural for them.

Sounds exactly like what Old Nick, if pressed on the issue, might say about Jack, right? Mill/Taylor would have none of it:

> I consider it presumption in anyone to pretend to decide what women are or are not, can or cannot be, by natural constitution. They have always hitherto been kept, as far as regards spontaneous development, in so unnatural a state, that their nature cannot but have been greatly distorted and disguised; and no one can safely pronounce that if women's nature were left to choose its direction as freely as men's, and if no artificial bent were attempted to be given to it except that required by the conditions of human society, and given to both sexes alike, there would be any material difference, or perhaps any difference at all, in the character and capacities which would unfold themselves.
>
> *(Mill, 2007, 178)*

In her magnum opus, *The Second Sex*, existentialist philosopher Simone de Beauvoir made essentially the same point: a woman's "wings are clipped and it is found deplorable that she cannot fly" (1952, 672). Mill/Taylor and de Beauvoir are employing what philosophers call counterfactual thinking, asking us to consider what women's accomplishments would be like *if*—contrary to established fact or practice—they had not been systematically deprived of access to the social and cultural means of proper development. The answer is supposed to be obvious to all but the most bloody-minded defenders of the traditional order.

Speaking of which, here's another example of the same sort of mischief. The Canadian psychologist and provocateur Jordan Peterson has suggested, apparently without irony, that social and gender hierarchies among humans are justified because lobster societies have such hierarchies and we share a common evolutionary ancestor with this creature. Inspired by this stunning new argument against equality, young men can now be found walking around with lobster T-shirts in support of Peterson and good old-fashioned—I mean *really* old-fashioned—hierarchies.

In addition to living in hierarchies, lobsters also occasionally eat other lobsters. Perhaps we should copy *this* behavior as well and become cannibals? At least, *occasional* ones since this is as far down that gastronomic path as our crustacean heroes have gone and, anyway, why forego filet mignon altogether? Like all defenders of the status-quo, Peterson cherry-picks his natural facts to support a social order he likes on entirely independent grounds, such as the fact that he is at the top of the current gender hierarchy and enjoys the view from there. His lame appeal to the lives of lobsters is clearly no better than the arguments of Aristotle's masters or Mill and Taylor's Victorian patriarchs. In fact, purely in terms of its persuasive power, I'd put it on a par with Old Nick's child-rearing philosophy.

Again, appeals to nature of the sort we have been investigating are almost always seeking to perpetuate some system of social power by showing that it is rooted in Nature. If the behavior or practice in question works in Nature then who are we to fight it? Talk about tilting against windmills! But Nature places very few specific constraints on our institutions and the values that partially define them. If we are inclined to point to Nature as a foundation for this or that practice or institution we should at the very least be aware that such arguments have historically been used quite often to put or keep certain groups—women, LGBTQ people, ethnic minorities, etc.—in subordinate social and political positions.

The basic error of all these theories is that they underestimate the extent to which Nature gets refracted through culture. Too often, when we think we are holding the mirror up to Nature we are really just holding it up to ourselves and valorizing the thing we find—but were already looking for—by labeling it 'natural.' In an intentional rebuke to this mode of thinking, the 18th–19th-century philosopher and politician Jeremy Bentham (1748–1832) once declared that the concept of rights was nonsense and that of natural rights "nonsense on stilts." That's going too far—as we'll see in Chapter 8, rights are much more important than Bentham supposes—but taken as a critique of any attempt to ground our values in Nature it's not terribly wide of the mark.

The Anthropocene

But Nature won't go away so easily, and in spite of what has been argued so far it is not yet obvious that we should show it the door. What Mill and Taylor say in the passage quoted earlier is actually rather complicated. They argue that if women were unshackled their true *nature* would shine forth and that the social shackles are, because of this, arbitrary and *unnatural*. Clearly, Mill and Taylor do not think we should abandon thinking about Nature authoritatively—as a foundation—only that we should be careful about such appeals.

For example, we might notice that ecosystems display a certain equilibrium of internal energy transfer, to the point where they look like semi-closed systems. One upshot of this evolutionary achievement is that such systems have found really efficient strategies of waste management, so that nutrients are cycled through the system, basically forever. Why not appeal to this as a model for our own attempts to manage waste? There's a whole field, called 'biomimicry,' devoted to searching for answers of this sort to many of our most intractable environmental and design problems.

I applaud biomimicry in general, but it is a mistake to think of it as an attempt to provide a foundation for our practices and values. And that's because we don't have access to anything fully natural in the first place. What we are seeking to mimic is already, in large part, a cultural production. This claim sounds strange, so in this section we'll pick it apart. Let's begin by taking a closer look at a new idea invading our world, that of the 'Anthropocene.' The concept of epochs is the creation of the field of geology, more specifically that of stratigraphy, the study of the layers making up the Earth's crust. We can understand major breaks in the history of the planet by looking at changes in layers of sedimentary rock. This is because when major changes occur on the surface of the planet, they are often recorded in these layers.

For example, there have been five major extinction events in the history of life on Earth. By 'major extinction event' we're talking about the demise of anywhere between 50% and 95% of species. When plants and animals die, their remains are laid down and compressed in sediment. The precise makeup of any layer of sediment provides proxy information about the mix of species that inhabited that place at that time. So in exposing a particular piece of Earth to view, cross-sectionally, stratigraphers might notice that there is a relatively sharp break at some point, a clear horizontal demarcation.

If this is the case, and the break can be attributed to an extinction event (though there are many other ways it can happen), a new historical designation may be called for. As we have seen, since the advent of agriculture some 12,000 years ago the Earth has been in the Holocene epoch. Now, however, geologists are in the process of deciding whether or not we have entered the Anthropocene. They have yet to decide the matter officially, but all signs point towards an affirmative answer.

It will be a momentous decision because it indicates that the human impact on basic planetary processes is so profound that it is discernable in the sedimentary makeup of the crust, along the lines of the process just outlined. Suppose our species meets its own demise within the next few hundred years and alien geologists visit our planet well after that. To say that we have entered the Anthropocene means that these geologists would be able to discern that sharp horizontal line in the rock-record. They might not realize that a single species was responsible for this alteration (they'd need a whole lot more archaeological information to determine this), just that the break was clear. What exactly would they find?

In 1986 the International Council of Scientific Unions established the International Geosphere-Biosphere Program (IGBP), whose task was to

> describe and understand the interactive physical, chemical and biological processes that regulate the total earth system, the unique environment it provides for life, the changes that are occurring in that system, and the manner in which these changes are influenced by human actions.
>
> (quoted in Angus, 2016, 30–31)

Guided by this mandate, stratigraphers are looking for what they term the Global Boundary Stratotype Section and Point (GBSSP), more colloquially known as the golden spike. This is that horizontal break in the sedimentary record we've been talking about, the historically novel section of material elements. So far, the best material candidates for the Anthropocene golden spike appear to be radionuclides associate with nuclear arms testing, plastics, carbon isotope patterns and industrial flyash. These markers would put the beginning of the new epoch at about 1945 (Zalasiewicz et al., 2017). This proposal is gaining assent among geologists.

However, some scientists understand the transition a bit more loosely. For them, among the key markers are the near exhaustion of fossil fuels in just a few generations, the transformation of half the planet's land surface, the uptick in nitrogen fixing, the appropriation of fresh water reserves by humans, increased concentrations of GHGs in the atmosphere, the alteration of coastal and marine habitats, the depletion of marine fisheries and the Sixth Mass Extinction (Angus, 2016, 35). This looser conception is most relevant to our purposes.

Because of its attention-focusing power, let's return to that last item: the Sixth Mass Extinction. In the previous chapter we saw that scientists have estimated that we are in the process of eliminating a million species from the world forever. One of the remarkable things about this is that we are accomplishing the job so quickly. Apart from the K-T extinction event which wiped out the dinosaurs some 66 million years ago and was relatively sudden, previous mass extinctions unfolded over millions of years. Ours is happening over a few brief centuries.

This event alone would be sufficient to change the sedimentary record, but the other items listed just above will have their effects on that record too. In the next chapter, I will talk more about the manner in which entry into the new epoch should affect our most basic sense of who or what we are as a species. We require a new historical self-understanding that incorporates the complex challenges of adapting to the climate crisis. We have, I'll therefore suggest, become *Homo crisis*, our very identity defined by the crisis-ridden turning point in Earth history we ourselves have unleashed. Here, however, I want to stick to the topic of foundations, and more specifically the sense in which Nature might still be able to provide a foundation, or model, for our values and institutions.

That ideal depends on our being able to identify Nature as something that is, to some degree, distinct from and independent of our cultural forms. And this is indeed the sense in which modern environmentalism has for the most part understood the concept of Nature. Nature, on this view, is that which is radically independent of human purposes. So far that is just a descriptive claim, a claim about the way some part of reality just *is*. But at least among environmentalists the descriptive claim often comes packaged with a prescriptive one, a claim about the way a part of reality *ought to be*. Here, the idea is that our ethical and political task is to conserve, preserve or restore the natural world, or some part of it, to the condition it was in before humans altered it. This notion, for example, is the philosophical inspiration behind the National Park systems in Canada and the US.

In a brilliant and provocative recent analysis, philosopher Steven Vogel (2016) has highlighted just how problematic this view is. The most general problem with it is that in the Anthropocene it simply does not look as though we can locate the line between nature and culture in the manner required to sustain it. To begin, there may never have been an independent Nature in the first place, at least not one that we could ever locate. This problem shows up in what is known as the historical baseline issue in conservationism. If we want to restore a land-scape to some previous state, which one do we choose? No matter how far back we go, we will find that the landscape of that time had been altered by humans.

For example, the indigenous peoples of North America altered their land-scapes profoundly, through both hunting practices and controlled burnings of forested areas. And even if establishing an historical baseline were not a problem, restoring nature to that state is clearly a human intervention into Nature. Re-introducing wolves into Yellowstone Park is a way of returning that ecosystem to some previous state—the one before this species was extirpated—but the scheme obviously has *us* written all over it.

The more general problem, for Vogel, is that the effects of human economic activity are seemingly everywhere. There are plastics in the bellies of whales, the shells of the tiny coccolithiphores making up the base of the marine food chain are softening because of ocean acidification, giant parcels of ground are sinking where huge aquifers have been depleted, methane clathrates deep in the oceans are warming up, northern permafrost is melting, we are homogeniz-ing the planet's collection of non-human species, GMOs are inside our bodies. Vogel invites us to try and name a place on the planet where our presence has not morphed reality to *some* degree, where our technologies or their byproducts have not penetrated. There are none.

Of course, much of this is due to anthropogenic climate change, a phenome-non that has altered every square centimeter of the planet at the chemical level. At the macro-level, there are more and more *mismatches* among species in ecosystems that had previously been symbiotically organized. Chicks hatch before their food sources have arrived on the scene. Plant and animal species are driven from the south into northern spaces, or from lowlands into alpine regions, to find a habitat more like the one they are used to, altering predator-prey balances. And so on.

Here's another way to make the basic point, one that returns us to the National Park ideal in North America. That ideal is pointedly *laissez-faire*. It is based on the thoroughly understandable idea that a big part of the environmental prob-lem has to do with our evident inability to leave anything alone. As soon as we find a bit of 'untouched' Nature, we want to inhabit it, transform it, mine its resources, etc. National parks seek to declare chunks of the world off-bounds to this sort of aggrandizing encroachment. Sometimes people are not allowed in at all, but even where they are allowed in it is as tourists only. This is what the environmental philosopher Ronald Sandler has called a "place-based" strategy of nature preservation. Just put up fences keeping the humans out—at least the ones bent on any kind of commercial development of the area—and let Nature do its thing.

That sounds great and it *was* a wonderful ideal. But what if the encroachments *seep through* the fences? Here's Sandler on this possibility:

> Place-based conservation strategies depend upon the relative stability of background climatic and ecological conditions. Global climate change disrupts that stability. To the extent that it does so in a particular location, place-based preservation strategies for the at-risk species that are there are less viable. They cannot preserve the species' form of life in their ecological context.
>
> *(2015, 359)*

Because some version of place-based preservationism has been the guiding ideal of environmentalism for the past 40 years, this is an immensely significant development. Climate change has fundamentally changed what 'Nature' means and can mean. If the old ideal is no longer tenable, then we need to think seriously about what it means to be an environmentalist now. But that puts the problem far too narrowly because the insight challenges how any of us, whether or not we identify as environmentalists, should think about that fundamentally odd triangulation of humanity, Nature and technology.

All of this can be summarized in two fundamental points. The first is that our technologies, or the effects of their applications, are now virtually everywhere. As the sci-fi writer Kim Stanley Robinson (2020) has put it, we exist not in Nature but in an "accidental megastructure." The second is that there is no walking this back. Any attempt to do so—assuming, for the sake of argument, that it is technically and economically feasible—will entangle us in historical baseline issues that are, as far as I can see, not rationally resolvable.

All we can therefore do is transform the ugly and unsustainable features of the accidental megastructure into something more conscientiously beautiful and sustainable. The central message of the Anthropocene is that the ways we will henceforth interact with *everything* else there is will be mediated by technology. To the extent that we are unable to escape nostalgia for untrammeled Nature, we will have failed to grasp this most uncanny aspect of our times.

Conclusion

Were we never to have stumbled into an ecological catastrophe, we'd still be faced with the task of learning how to live in a world deprived of its traditional foundations. There's no mention of the environment in anything Nietzsche says, although Heidegger has become the inspiration for a certain brand of philosophical environmentalism. Modern science all by itself has spun the world free of its moorings. Early in the 20th-century, the German sociologist Max Weber coined a term for the modern scientific interruption, 'disenchantment' (in German, *Entzauberung*, literally 'demagication').

Talk of enchantment, or its absence, puts the home back in the house, even if only as an unreachable ideal. It indicates that we are not speaking here of a mere structure but a habitation or dwelling, a construct expressing our values and aspirations. A keep. That is what you see embodied in great buildings like Notre Dame, and what made the recent fire there so shocking.

As I have been arguing, those (like Pinker) who ballyhoo the relentless progress and forward motion that define the modern period often downplay this other side of modernity's coin. But Weber too knew nothing of ecological crisis, and so the disenchantment *we* are now faced with is layered or double-barrelled (pick your metaphor). For whereas those early moderns could still have looked to Nature as a model, perhaps learning how to do this in a more responsible way than was often the case (as Mill/Taylor did), we do not have even that luxury. Our disenchantment is deeper than theirs because we have lost Nature as well as the God of the philosophers, the one *reason* purports to reveal. We are walking on metaphysical sand.

Still, I hope you have not come away from these chapters entirely crestfallen. Climate-induced ravages aside, there remains a beautiful biosphere to behold, a teemingly rich life-world worth fighting for. After all, is birdsong less apt to lift your soul just because spring has arrived weeks earlier than normal thanks to *our* heating of the planet? Are you less inclined to admire the curl and crash of ocean waves whose kinetic energy *we* have amped up? I doubt it. Nature is still chock-full of sublime and often mysterious wonders. It's a model for our practices in limited contexts, as with biomimicry. That should provide some solace, even if Nature can never again be a foundation for our values. But now, alas, it's time to look at the ravages more closely. Time to face the crisis head on.

References

Angus, I. (2016). *Facing the Anthropocene: Fossil Capitalism and the Crisis of the Earth System*. New York: Monthly Review Press.

Baier, A. (1994). *Moral Prejudices: Essays on Ethics*. Cambridge, MA: Harvard University Press.

de Beauvoir, S. (1952). *The Second Sex*. New York: Knopf.

Locke, J. (1980). *Second Treatise of Government*. Indianapolis: Hackett Publishing.

Macpherson, C.B. (1962). *The Political Theory of Possessive Individualism*. Oxford: Oxford University Press.

Mill, J.S. (2007). *The Subjection of Women*. New York: Penguin.

O'Connor, F. (1979). *The Habit of Being*. New York: Farrar, Strauss and Giroux.

Paine, T. (1961). *The Rights of Man*. New York: Dolphin.

Pascal, B. (2008). *Pensées and Other Writings*. Oxford: Oxford University Press.

Robinson, K.S. (2020). "Imagining a Flooded Planet." *The Brooklyn Rail*. Retrieved from: https://brooklynrail.org/special/River_Rail_Colby/river-rail/Christopher-Walker-with-Kim-Stanley-Robinson. Accessed February 12, 2020.

Sandler, R. (2015). "Global Climate Change and Species Preservation." In *Environmental Ethics for Canadians*, Byron Williston (ed.). Don Mills: Oxford University Press, 356–365.

Sekar, S. (October, 2015). "A Few Species of Ant Are Pirates that Enslave Other Ants." *BBC Earth*. Retrieved from: www.bbc.com/earth/story/20151028-a-few-species-of-ant-are-pirates-that-enslave-other-ants. Accessed March 29, 2019.

Vogel, S. (2016). *Thinking Like a Mall: Environmental Philosophy After the End of Nature*. Cambridge, MA: The MIT Press.

Wollheben, P. (2016). *The Hidden Life of Trees*. Vancouver: Greystone.

Zalasiewicz, J., et al. (2017). "The Working Groups on the Anthropocene: Summary of Evidence and Interim Recommendations." *Anthropocene* 19, 55–60.

3

THE SHAPE OF OUR CRISIS

In early April, 2019, three climbers were killed in Banff National Park in Alberta, Canada. The most striking thing about this terrible event, to me at least, is that the three climbers—an American and two Austrians—were world-renowned experts in their sport. The peak they were attempting to scale was Howse Peak, one of the world's most forbidding climbing challenges. A Parks Canada official noted that the face they were attempting "is remote and an exceptionally difficult objective, with mixed ice and rock routes requiring advanced alpine mountaineering skills" (Flynn, 2019).

Maybe these elite athletes could have made the climb successfully if the snow hadn't been unstable. But then again, in that place at this time of year negotiating tricky snow conditions is part of the job of climbing, isn't it? Why were three of the world's most experienced alpinists, presumably equipped with the best gear money can buy, not prepared for *these* snow conditions? Is the snowpack behaving weirdly now? If so, is this related to climate change? The answer to the last two questions seems to be yes. Snow avalanches are complex phenomena. Johan Gaume is a researcher at the École Polytechnique Fédérale de Lausanne in Switzerland. His work focuses on the way the snowpack is changing dangerously due to climatic alterations. His simulations indicate that climate change is creating ideal conditions for increases in snowpack "instability," leading to more so-called wet snow avalanches (Emory, 2019).

There's a larger question here, one about the relation between climate change and individual weather events. Is climate change responsible for such events? If so, can we demonstrate this? This question, the so-called attribution problem, surfaces whenever climate catastrophe strikes. The NYU philosopher Dale Jamieson has a useful analogy explaining why it is in one sense a misguided question. A baseball player's batting average gives us a pretty good idea of how many hits he's likely to get per game or over the course of, say, a week or month of games. But the average does not *cause* any of the hits. Statistics don't play that kind of role in the material world. Climate science also reports mere statistical averages. How could *they* intervene in the weathery world? To suppose they could is to make a basic category error.

True enough, but the science of event attribution *is* making rapid progress. For example, in the coming decades we can expect a marked increase in the intensity or frequency of extreme weather events—floods, hurricanes, typhoons, mudslides, heat waves, forest and bush fires, etc. Attribution science can help us distinguish anthropogenic causes from natural variability with respect to these events without committing the crude fallacy Jamieson points to. It shows how a supercharged climate system makes individual disasters more likely than they would otherwise have been. In repeating ad nauseam that this or that weather event cannot be linked to our activities, climate deniers have simply failed to absorb the latest science. This failure should cause us reflects more deeply on the chaotic climatic world we have made and are making.

That last sentence expresses a telling paradox: we are makers of chaos. This chapter is all about that uncanny role we have assumed in the world outside our windows. The story of the three alpinists is a sad illustration of a more general truth. If we suppose that anthropogenic climate change made that avalanche more likely—and this does not appear at all far-fetched—then *we* are causally responsible for this disaster. Our fingerprints are all over it, and yet it is an event of overwhelmingly powerful destructive force. We have very little chance of controlling happenings like this. That, in a nutshell, is life in the new climatic regime we have wrought: causal responsibility for a world largely out of our control.

We'll spend some time looking at this aspect of the new epoch, and ask what sorts of challenges it will present to us. The crisis in which I'm mostly interested here, however, is the crisis of values that could beset us in a world that is pitched into situations of extreme resource scarcity due to climate change. The age of comfortable illusions—the Holocene, in a word—is behind us.

Homo crisis

Nobody has any idea how long the Anthropocene will last or even what it will look like a few centuries hence, but there is no disputing that the signature event of its early years is anthropogenic climate change. I've been making frequent references to this phenomenon, but so far these have been mostly data-free. Now it's time to ground our understanding of it with more pointed references to the relevant science. Since roughly 1990, after five (and soon to be six) massive reports from the IPCC, it has become crystal clear that our current trajectory is leading us to climate disaster. The unintended consequences of our land use practices, reliance on fossil fuels as our primary energy source and heavy use of ruminant agricultural animals (especially cows) are now having measurably dramatic effects in the Earth system and, by extension, on our lives and livelihoods.

There are now more books, journal articles, YouTube videos, blogposts, newspaper articles, and social media rants about this topic than I can count, and the last thing I want to do here is throw another heated analysis on this pile and leave it to molder with all the others. So, I'm not going to waste time showing that climate change is real, that climate change deniers are demonstrably unserious people, that we are causing the Sixth Mass Extinction event in the history of life on Earth and so on. Anyone genuinely inclined to doubt any of this should consult the science, beginning with the latest report from the IPCC. The adjective that body uses to describe the basic facts of anthropogenic climate change says it all: the evidence for this, they say, is "unequivocal" (IPCC, 2014). It's time for our understanding to move on.

One of the latest salvos from scientists is aimed directly at humanity's inertia on the climate file, a shocking report about the state of the climate released by the UN late in 2018. It focuses on the desirability and feasibility of our keeping the global average temperature increase to 1.5° Celsius by the end of the century, relative to the pre-industrial baseline. The report is shocking because it shows, among other things, that the difference between this target and the slightly higher one of 2° Celsius would be dramatic as regards their respective effects on global weather events. For example, if we are able to cleave to the 1.5° target—the aspiration of the Paris Climate Accord—the number of people experiencing water stress could be cut in half compared to the higher target. Similar figures apply to the issues of food scarcity, sea-level rise and so on (Watts, 2018).

So let's just stick to that lower target, right? Sounds great, but I'm afraid it's simply not going to happen. UN reports like this always come leavened with what their authors take to be just the right amount of gumption-inducing hope that we can turn the ship around in time to avert disaster. Even so, the report claims that unless we achieve a significant alteration in the global energy market *by 2030* we won't make our target. Because we have already added more than 1° Celsius this is an extraordinarily tall order, meaning that we are likely headed for 2° Celsius. Beyond that things are less certain. If we blow past 2° Celsius, there's a very good chance that we are heading for an end of century warming 3–4° Celsius higher than that pre-industrial baseline.

If we walk into it unprepared, confident that we can get by just by battening down the hatches, that is a civilization-ending result. Billions of people will die prematurely, regional wars generated by resource scarcity will break out (possibly among countries with nuclear weapons, like India and Pakistan), climate refugees will pour across international borders, fascist demagoguery and outright warlordism will rise dramatically in many parts of the world, the biosphere will lose half or more of extant species, etc. Should I go on? I'm no fan of disaster porn, so I don't think I will.

Again, the literature on all of this—both popular and academic—is easy to find. For popular summaries of the catastrophes likely awaiting us, start with Mark Lynas' 2008 *Six Degrees: Life on a Hotter Planet*, then move to David

Wallace-Wells' 2019 *The Uninhabitable Earth: Life After Warming*, a scientifically updated version of the same thing. These are objectively terrifying analyses.

Now that we know what will happen if we don't take appropriate action, why am I so confident that we are *not* going to do so? Why the sour face? Let's look at some more numbers, this time concerning likely patterns of energy use now and in the near-term future. According to the International Energy Agency (IEA), the gold standard for global energy trends, energy demand grew globally by 2.3% in 2018, with China, India and the United States accounting for most of that growth. Fully 70% of this demand was met by fossil fuels, leading to a 1.7% increase in greenhouse gas emissions. Coal, that unapologetically carbon-rich energy source, *still* accounts for 30% of energy-related CO_2 emissions (IEA, 2019).

That's 2018, but what about trends going forward? Remember, the UN's 2018 report gives us until 2030, so maybe we will have weaned ourselves off of fossil fuels by then. I'm skeptical. COVID-19 has slowed global energy use down substantially, even knocking some oil stocks down into negative territory, but that's likely a blip because we have no replacement for this commodity in the quantities required by the global economy as currently structured. When the virus is defeated, these stocks and this commodity will therefore probably roar back.

Thus BP—which paints itself, risibly, as an environmentally conscientious oil company—is (unfortunately) seemingly correct to argue that "significant levels of continued investment in new oil will be required to meet oil demand in 2040" (BP, 2019). Companies like BP are counting on governments to continue subsidizing their operations in a way that helps realize this outcome (including through COVID-19 stimulus packages). Of course, subsidies to fossil fuel companies are the diametric opposite of a carbon tax, one of the policies required to decarbonize the economy in line with the UN's recommendations. Subsidies entrench the commodity in the market, while the tax seeks to push it out (ideally in a progressive way).

And yet governments still subsidize the fossil fuel industry to the truly jaw-dropping tune of about $5 trillion/year (IMF, 2019). Among other things, that is money used to build the infrastructure of the fossil fuel industry: pipelines, refineries, R&D on exploration technologies and so on. Let's be blunt, please: these companies and governments are investing in a commodity they expect to be around for decades, a *robust* part of the global energy regime well beyond 2030. We cannot have it both ways: we cannot *both* refuse to impose hard constraints on fossil fuel companies seeking to extract and burn this commodity *and* expect to avert disaster.

Even if we could rein in the subsidies to some extent, the task facing us would be enormous. With the best will in the world we will have squandered the 1.5° budget with just nine more years of greenhouse gas emissions at current levels. If we had started the task of serious mitigation in 2000, we could have met the target by reducing our emissions by 4%/year. That would have been hard, but not impossible. But because we have done nothing, according to a recent UN report we are now faced with the task of reducing emissions by 7.6%/year. The report adds somberly that "collective ambition must increase more than fivefold over current levels to deliver the cuts needed over the next decade to meet the 1.5° target" (UNFCCC, 2019).

Given the improbability of emissions cuts reaching this level starting *right now*, many believe that the only way to reach the target is by coupling slightly less aggressive emissions reductions with so-called negative emissions. This involves sucking existing carbon out of the sky. There's been a bit of a global buzz lately about massive afforestation projects, for example. We hear reports about national governments, including that of the US, committing to planting billions of trees. If these trees can be burned for energy, and the resulting carbon buried successfully, it is, so goes the argument, a win-win policy. This is "bioenergy with carbon capture and storage." It sounds impressive, but it is almost entirely notional at the moment. According to a 2017 study published in the journal *Earth's Future*, this technology's "carbon sequestration potentials and possible side effects still remain to be studied in depth" (Boysen et al., 2017).

Because of all of this we are now *committed* to some not insignificant level of climate disaster. To help contextualize this geologically, compare what climatologists have discovered about the differences between two epochs, the Holocene and the Pleistocene, with respect to carbon concentrations found in Greenland ice cores over this long period (the past 100,000 years). If we take carbon concentrations as proxy evidence for average temperatures, the first thing to notice is the relative coolness of the pre-Holocene period. Just as importantly, Pleistocene temperatures fluctuated wildly compared to their relative stability throughout the Holocene (Ganopolski and Rahmstorf, 2001). A wild sawtooth lasting for 88,000 years followed by a 12,000-year solid line.

In his magnificent research on the prehistoric climate, William J. Burroughs (2005) describes the difficulties our ancestors must have had negotiating the crazy fluctuations in climate that mark most of the Pleistocene epoch. In a lovely phrase, Burroughs refers to the emergence of the Holocene period from the Pleistocene as "the end of the reign of chaos." Climate defined the parameters of life for humans prior to the Holocene in a brutally deterministic fashion, in particular by dictating strict limits on population. In the midst of ice ages, food sources can become suddenly scarce so you had better make sure there are not too many mouths to feed. Besides, we are talking largely about nomads who have to travel, sometimes great distances over harsh landscapes, to find food. All by itself, that places restrictions on group size. Only sedentary peoples can afford population explosions, other things being equal.

The Holocene is a time marked by a vastly expanding human population, the emergence of hierarchically organized social structures, government bureaucracies, sprawling cities, complex religions and exponential growth in the technologies of agriculture and warfare. All of this could happen because by and large humanity could, finally, forget about the damn weather! That's the Holocene:

the epoch in which the weather became an item about which to chat casually around the ancient water cooler rather than a fearsome beast always poised to snatch away the most vulnerable members of your group. Well, guess what? The beast is back. Imagine the timeline of that graph projected out to the next 1,000 years or so. We can't say with certainty what it will look like but what we *can* say about it is not particularly encouraging.

Most obviously, average temperatures are going to be above the Holocene line (in fact they already are). But because of the possibility of abrupt, non-linear changes in the climate system they will also be significantly more variable. The turbulence may not turn out to be as dramatic as that of the Pleistocene, but it does not need to be in order to put enormous pressure on the infrastructure of our current systems. If we don't take measures to control global population our numbers will likely peak at about 10 billion by mid-century. This is a growth trend entirely dependent on that relatively placid Holocene temperature line. But if the line is going to move away from placid solidity to something more saw-toothy, much of the infrastructure of our civilization is going to become more or less instantly obsolete.

Because of all of this, adaptation to the effects of climate change in the early decades, or even centuries, of the Anthropocene is going to redefine chaos. What we face will make the travails of our ice-age ancestors look like a walk in the park. Clearly, we need to get a handle on this, and fast. But that is going to be exceedingly difficult, because the Earth system seems to be spinning out of control. This brings us to the central paradox of the Anthropocene. We mustn't conflate our ineliminable *presence* in the Earth system with the idea that we have *control* over it. The example of the deadly avalanche in Banff illustrates the distinction nicely. There's a meaningful sense in which we caused that event, but we can't control events like it.

Since the distinction is very important, here's another—much less tragic—illustration of it. Humans have been breeding dogs for around 20,000 years. That's a really, really long history of biological engineering, and yet it does not mean we have full control over the behaviors of these animals. My dog, an English Cocker Spaniel named Ridley, recently got his teeth into a one kilogram bag of brown sugar dangling too close to the edge of the kitchen counter. He ate every last mote of the sugar as well as the plastic bag containing it. An hour later, he vomited up a pasty beige slurry all over the kitchen floor.

Never one to pass on anything smelling even remotely ingestible, he then proceeded to eat the vomit, causing him to throw it all up again just a few minutes later. I can't say for how long this disgusting cycle would have gone on had I not intervened and put a stop to it. From a design perspective, characterizing the dog's behavior here as sub-optimal seems an understatement. I really would have thought that after some 20 millennia of concerted canine engineering we might have found a way to breed eating-your-own-puke out of this species. Evidently not.

On the other hand, Ridley's a great dog overall. He's cute and cuddly, playful, really social and fantastic with kids. In other words, we've made a great deal of *progress* selectively engineering animals like this out of their comparatively uncuddly and anti-social lupine forebears. Conclusion? The modern dog is both highly engineered and relatively uncontrollable. Writ very large, that's also the best way to think about the Anthropocene. It's the uncontrollable aspect of 'nature' that should worry us because, unlike Ridley, it doesn't care if we live or die.

In spite of the surfeit of worried analysis we now encounter about climate change not many have grasped that because of it we humans have now made *ourselves* into something brand new: *Homo crisis*. We have become, constitutively, a species whose place in history is going to be defined by adaptation to ongoing disaster. *Homo crisis* is, obviously, a species-level identity ascription. But to the extent we embrace it, it will come to color the way we understand the lower-level identifications making us who we are: international (or cosmopolitan), regional and bioregional, national and sub-national, religious, familial, romantic and more. The climate crisis is in the early stages of assimilating all these other, thicker modes of communal being to its own organizational prerogatives. It is becoming totalizing.

Given this, here's a crisp way to state where this book's argument is at the moment, one that focuses on the relation between crisis and foundations, and remembering that foundations provide ways of coping with crisis because they stabilize our values. There are three broad possibilities. The first is that the foundations are solid and there is no crisis. These are happy times for humanity. The second is that there is a crisis but the foundations are still solid. This is trying, but it's a situation that will, with a bit of luck, likely pass. The third is that there is a crisis and the foundations are gone. That's the lot of *Homo crisis,* and it is a profoundly disorienting situation to be in. In fact, it threatens to undermine our most cherished values, the pillars of democratic life.

Moral inclusivity

As we have seen in chapter 1, Pinker is to be commended for demonstrating that in the post-Enlightenment period humans have made significant progress in reducing violence, increasing life expectancy, reducing poverty and child mortality and extending education to women. Yes, there's loads to be done on all these fronts, and there has been serious backsliding on them in this historical period, but it's not rational to deny the broadly progressive trend in these categories of human well-being. Let's dig a little deeper into this claim and ask what we mean by "progress".

To begin, we often assume that the pied piper of technological progress brings moral progress irresistibly along with it. This, however, is clearly false. For example, the world is currently in the most technologically advanced state it has ever been in at the same time as inequality has soared. According to the latest data from Oxfam International (2020), there are just over 2,000 billionaires in the world. The collective wealth of this group exceeds that of 60% of the rest of humanity, fully 4.6 billion people. This is a moral scandal, unfolding right in the middle of an ever-expanding technological world of wonders. Clearly, we must analyze the two forms of progress separately.

Technological progress is comparatively easy to understand. It happens when some new device or technique of information-organization helps us negotiate our way through the world more effectively. What about moral progress? This is a more difficult notion to grasp, partly because it is historically variable. We require a definition of it that is (a) specific to our historical self-understanding in the modern democratic age and (b) as ecumenical as possible within this historical time-frame.

Let's therefore say that we make progress, morally speaking, when we extend the circle of moral considerability outward to encompass a class or category of beings we had previously considered beyond the moral pale. We don't have permission to treat anything of moral considerability any way we like, and this is because such things have interests that count. Moral progress is thus a matter of increasing inclusivity, moral regress a matter of increasing exclusivity. As a description of what underlies the democratic ethos most, I think, would agree with this idea, at least in broad strokes.

If so, the next step in the argument asks us to notice that the circle of moral considerability has without doubt been getting wider and wider. We once believed that children were beyond the pale, that members of other races were and that women were. But in many places in the world—not all, to be sure, but at least in those societies whose institutions are broadly liberal-democratic—members of these groups are now inside the circle. Even in the world's bastions of progressive values, however, this broad outward movement is not something to be taken for granted. Not in the age of climate crisis.

Philosophers Allen Buchanan and Russell Powell adopt a version of the view of moral progress just described. They also argue that the moral inclusivity/exclusivity distinction maps onto a distinction between favorable and unfavorable environmental conditions. Moral inclusivity, they claim, is a "luxury good" that becomes available only when environmental conditions allow human groups to bring in outsiders. The flip side of inclusivity is the urge to exclude, expel, demonize and even kill those on the outside. Buchanan and Powell write:

> Exclusivist moral response is a conditionally expressed trait that develops only when cues that were in the past reliably correlated with outgroup predation, exploitation, competition for resources and disease transmission are detected.
>
> *(2018, 15)*

If we think unflinchingly about the way our world is now, and will likely be in the decades to come, must we not admit that these "cues" are likely to become ever more detectable? To take just one example, climate change is going to expand the range of disease vectors like malaria-carrying mosquitoes *at the same time* as waves of people, fleeing extreme heat and rising seas or seeking out sources of fresh water, are compelled to migrate across international borders.

That, and related scenarios, are recipes for a potentially terrifying resurgence of moral exclusivity. It follows that protection of the ongoing project of moral inclusivity requires doing everything we can to cultivate the material conditions required for it to flourish. But, as we saw in the previous chapter, because we are now in the Anthropocene, the notion of 'material conditions' now has a special meaning. It entails that we must double down on the *technological enhancement* of the spaces we inhabit.

That is, for us the environmental preconditions for the flourishing of inclusive values, their material scaffolding, must be *constructed*. We do not have the option of letting nature be, both because there is no such nature and, more importantly, because this hands-off approach rests implicitly on the dangerously outmoded belief that 'nature' is our friend. It is not. I don't blame 'nature' for this post–Holocene about-face. After all, to reinvoke Wally Broecker's terminology, we're the ones who got it so ornery in the first place. The unavoidable upshot is that if we build badly or timidly now, we will also likely backslide badly in the moral sense.

Though the task sounds paradoxical, we must construct a Holocene-like world—minus the fossil fuels, of course—even as the sun sets on that epoch. Such furious and conscientious building guarantees us nothing but, as far as I can see, there is no more morally defensible way forward. I will expand on this technological imperative in Chapter 9, taking special care not to fetishize it in a manner that forgets about the distinction between the two kinds of progress elaborated in this section. The point here is that in the Anthropocene getting the *preconditions* for moral progress right is an ineliminably technological enterprise. This is because technology has now engulfed the world, whether we like it or not.

It is tempting to dismiss this as self-congratulatory techno-optimism, but that is not at all what it is. Since I'm apt to be misunderstood on this point, let me close this section by saying why it's a mistaken interpretation of what I'm claiming. Even if the main lesson of the Anthropocene is that the world is already permeated by technology, and that we cannot somehow negate this reality, the question remains as to what specific sets of values—ethical, aesthetic, spiritual, political, economic—will guide our design projects. But there's no reason to believe that this must be in the direction of homogenizing the natural world *solely* to better serve human purposes. There's a confusion at the heart of many philosophical attempts to grasp this point.

For instance, in an illuminating dialogic account of the meaning of 'nature' in the new epoch, the German philosopher Michael Hampe has one of his characters say this:

> [T]he technological world is a place where humans, as you'll admit, encounter only each other and their own needs. . . . . [T]echnological

development is driven solely by human self-reflection and the question of how these needs can be satisfied in an objective world.

(2015, 91)

This is an understanding of technology's essence that goes back to Heidegger, for whom technology transforms everything in the world, including humans themselves, into what he calls "standing-reserve." As Heidegger puts it, when we conceive of the world as standing-reserve, "everywhere everything is ordered to stand by, to be immediately at hand, indeed to stand there just so it may be on call for a further ordering" (1993, 332).

Hampe's character and Heidegger assume that technology can be used to advance only human ends. That's why in a fully technologized world we would encounter only ourselves. But the inference is dubious. Techno-narcissism is a very real danger, but it is by no means an inevitable concomitant of the recommendation to fortify the world technologically. For, to take just one possibility, we can intervene in the lives of organisms and ecosystems in order to enhance *their* capacity to survive the ravages of climate change, with no reference whatsoever to *our* interests. Not all technological manipulations of the natural world need be aimed at making it better and safer for us, though it must be said that most of them at the moment certainly are given the uncontrolled way we have spread ourselves over the planet.

As my earlier comments about bioenergy with carbon capture and storage should make clear, I believe there can be sound reasons to reject any particular technological scheme. And I don't think those reasons can all be set out in terms of narrow cost-benefit analysis or the push to convert everything to standing-reserve. One of the problems with the bioenergy scheme, for instance, is that because of their "space-consuming properties" biomass plantations would likely fail to make up for insufficiencies in emissions reductions *without* also compromising "biosphere functioning."

The devil's in the details, of course, but that's surely a strong *prima facie* reason to be wary of the proposal or even to reject it altogether: *it poses a threat to the biosphere.* Some technologies will do this, others will not. But let's be clear about our choices here. It's not technology versus no technology, but good versus bad technology, where 'good' and 'bad' stand in for the whole cluster of values isolated earlier. Wise adaptation consists largely in discriminating among these types of cases.

I will elaborate on the complex relation between biodiversity protection and enhancement of the technosphere's design in Chapters 7 and 8. For the moment, I want to concentrate on the specifically human prospect. What we are now realizing is just how precarious the luxury good of inclusive morality is. It is the beating heart of the ongoing project of democratization, but its continued viability is not guaranteed by God, Nature or any other foundation. And now we have, all by ourselves, begun to undermine the material conditions required to host it. To help us get a grip on this, in the next section of the chapter I'm going to describe two general ways in which the value of inclusivity is going to come under pressure in the new epoch.

Two threats to inclusivity

The first and most obvious threat is that inclusivity will meet its opposite, exclusivity, more often and in more dramatic confrontations than is now the case. I wish this were as abstract a possibility as my wording here suggests. But we all know what I'm talking about: the way perceived scarcity can so easily become a trigger for toxic ethno-nationalism. For example, in November, 2019, the German city of Dresden declared a "Nazi emergency" (*Nazinotstand*). A local councilor defending the declaration noted that the rise in far-Right extremism all across Germany is a clear threat to "the open democratic society" (BBC, 2019).

One of the effects of climate change will be a sharp rise in climate refugees, and the far-Right is defined by its anti-immigrant animus. Assuming Buchanan and Powell are correct about the correlation between moral exclusivity and environmental degradation, we should expect a global spread of such worldviews in the coming decades. Xenophobic ethno-nationalism is evidently a force to be reckoned with. With respect to the clash of values—inclusivist versus exclusivist—this is pretty easy to analyze. Although a certain leader of the free world has opined that in cases like this there are always "very fine people on both sides," the truth is simpler.

This brings us to the second general threat, which is more complex. I'll set it up by pointing out that moral values usually come in packages. In fact, *webs* is probably a better metaphor. An exercise I like to do with my moral philosophy students is ask them to examine their values and see whether or not there are any conflicts among them. There usually are. For example, most people value freedom of speech, but some also think that institutions like universities have an obligation to protect other values such as diversity and equity. All decent people think of these three values—freedom, diversity and equity—as positive things.

However, sometimes they can come into conflict with one another, for instance when a university student group invites a white supremacist to campus to spread her word. It is not hyperbolic to say that this kind of speaker threatens the values of diversity and equity by inciting others to treat members of disadvantaged or marginalized groups in a disrespectful manner. The issue extends well beyond the confines of university campuses, and is not a case of (allegedly) entitled millennials being too emotionally brittle to entertain views contrary to their own. This is the uncharitable spin the Right puts on events like these. It's more accurate to point out that this is a value conflict: it's free speech *versus* inclusivity and respect.

My object here is to not to weigh in on this particular issue one way or the other, only to emphasize the seemingly ineliminable fact of conflict among positive values it illustrates. Most of us don't like the tension or even contradiction such conflict produces, so we seek ways to minimize it. This will usually involve reinterpreting one of the contested values in a way that reduces or eliminates the conflict between it and the others, and then trying to get others to accept our re-definitions.

This is all rational enough and something that should be encouraged in pluralistic societies. When these debates heat up on my campus, I try to remind people that confrontations like this are the very lifeblood of a free society. This can have the sobering effect of getting people to relax a bit, to stop thinking about social conflict as a *problem* to be fixed or avoided altogether. I need to say this here because of what I'm about to argue concerning the value conflicts we are inviting with unchecked climate change. There's a world of difference between garden-variety liberal-democratic value conflict and the sorts of tragic conflicts we seem headed for.

So let's get back to the issue at hand, the second threat. If values come in webs, then so does the value of inclusivity. For example, I assume that most people who value inclusivity also subscribe to a lot of other related values. Depending on the person, this list could be quite long, but here are a few notable items that are likely on it. The first and arguably most important is *toleration*, the ability to live with and even celebrate difference.

Then there's *compassion*, the willingness to put yourself in the shoes of other people and consider how your decisions make them feel. *Justice* is also a value, one that asks us to distribute the benefits and burdens of our decisions regarding the allocation of scarce resources in a way that is equitable across a population. And while we're at it, let's add *open-mindedness* to this list, the willingness to challenge your own views about the way the world works.

I'm happy to entertain other candidates for this list. These might include psychological and physical health, what the political philosopher John Rawls calls the social bases of self-respect, and just plain staying alive. Because we are talking about an internally coherent web, each of the values can contribute to or enhance all or some of the others. The open-minded are generally also tolerant, for example. Like philosopher Joseph Heath (2002) I'd even call efficiency a value and put it in my ideal web. As long as our political goals have been genuinely informed by the broad value of inclusivity, it makes sense to pursue the most efficient means to their fulfilment. Since it trains our focus resolutely on getting the outcomes we want given the scarce inputs available to us, acting efficiently makes it more likely that the goals will actually be reached. Efficiency in the service of good goals is itself a good.

What is really worrisome about the climate crisis is that it will force on us unprecedentedly hard choices, tragic choices, *among these positive values*. This is the essence of the second broad threat to the value of inclusivity. If the first threat is that inclusivity will meet its diametric other, exclusivity, this is about positive elements of the web coming into mutual conflict to a degree that will shake us to the core. The best way to describe this sort of outcome is to point to hypothetical scenarios in which it could emerge. Here are three.

The first two scenarios have to do with how we deal with acute strains on national health care systems caused by severe material scarcity and/or disease outbreaks. There's nothing merely theoretical about this. Because of COVID-19 many countries now have first-hand experience of it. Triage—saving only a *subset* of a population of desperate people—is a go-to coping strategy in these circumstances. How would we decide who lives and who dies? Anyway we like, in theory. One possibility is to resort to considerations of efficiency. Where resources are scarce, we will in this case save those who are, as the philosopher Catriona McKinnon puts it, "efficient convertors of resources into life" (2011, 120). Favoring the young over the old is one expression of this strategy. It might be an obvious solution in many cases.

But what if there's a conflict for scarce healthcare resources between two or more people of roughly the same age? Do efficiency considerations dictate privileging the most physically fit? Unfortunately, it's pretty easy to see that those who are fittest in this sense might also not be the most deserving in other respects. They happen to be healthier, but health is not a moral virtue. In fact, since health is strongly correlated with wealth, this way of doing triage might just end up favoring the already rich and powerful. That does not sound fair. This form of triage can therefore pit efficiency and survival against equity.

The second scenario is a survival lottery. It's just what it sounds like: we draw lots to see who lives and who dies. It sounds harsher than triage, but everything depends on how we think about the values involved. In fact, so long as the lottery is not rigged, it can be a much fairer way to make life and death decisions than triage is. Everybody has the same shot at survival. So we've taken care of equity or fairness, but the survival lottery clearly pits *those* values against compassion (Williston, 2015, 144–147). Can you imagine facing someone, especially a loved one, who has been consigned to death or exile just because she drew the short straw?

Both triage and the survival lottery involve innocent people dying. This is why I characterize them as clashes among our positive values rather than between those values and something alien to them.

The third and final scenario is like this but is also different in important ways. It concerns the likely rise of suicides. Suicide is not often seen a matter of conflict among positive values, but this description of it is apt. A value most of us hold dearly, self-preservation, must be abandoned because it conflicts with the value of living a life free of physical and/or psychological misery. Here's a vivid historical analogy of the kind of thing that worries me. In the middle decades of the 17th-century, the world experienced its own (non-anthropogenic) climate crisis.

This period is known as the Little Ice Age. Relative to the period 1000–2000, there was a drop in global average temperature of -0.6° Celsius. The historian Geoffrey Parker has written a massive book about this period. The general picture he paints is of repeated climate catastrophes exacerbated by stupid policy decisions on the part of national governments (with the notable exception of Japan's). The ultimate effect of all this was a significant uptick in mini-wars, revolutions, rebellions, witch-burnings, etc.

Also suicides. Parker cites a local account of suicides in Shandong, China: "Everyday one would hear that someone had hanged himself from a beam and killed himself. Others, at intervals, cut their throats or threw themselves into the river" (2013, epigraph). That's a grisly and heartbreaking reaction to the effects of a temperature anomaly that is small relative to the + 3-4° Celsius anomaly we are inviting. In a study about the connection between suicide and rising temperatures in the US and Mexico, scientists have shown that on our current path of warming this region could expect to see an increase in climate-related suicides of up to 40,000 people by 2050. That puts climate change roughly on a par with economic recession as causes of suicide (Burke et al., 2018).

Two final points before we move on. First, I want to emphasize a point made in the Introduction, namely that in the age of climate change crises will compound and cascade (Dufresne, 2019). For example, can you imagine what will happen to a country coping with a large-scale disease outbreak like COVID-19 *and* severe flooding on its southern coastline *and* a surge of migrants from beyond its eastern border? Should such crises become sufficiently widespread and serially entrenched it is not difficult to imagine them destabilizing whole political systems. This is what we must prepare for psychologically, politically and technologically. Which brings me to the second point.

We must understand the claims I am making about value clashes in the larger context of technological decision-making. Adapting to the climate crisis will require thinking about how we design all the parts of inhabited, and inhabitable, space. We will need to decide on the viability and advisability of geoengineering, synthetic biology, floating cities, artificial coral reefs, farm towers, gene and seed banks, expanded nuclear power capability and much more. The value clashes just described will arise, in varying degrees of tragedy and trauma, largely as a result of the ways we reconfigure space technologically. But because of cascading and compounding crises all of this building will create winners and losers. Space is irreducibly social and decisions about its design are therefore necessarily political. So it's time to say more about the political aspect of our situation, which I'll set up with a mostly friendly look at the Anthropocene naming wars.

The politics of the Anthropocene

Who exactly is the 'we' to whom I keep referring? This question is more difficult to ignore than you might think. While at the gym yesterday I was blasting Metallica through my headphones. A few minutes ago, I had a sensation of ringing in my ears which I immediately attributed to the probably excessive volume at which I was playing the music (Metallica, of course, *must* be listened to at this level, so it's not like I had a choice in the matter). However, it turns out that I was hearing the howls of protest from all those academics down the hall from my office who will not stand for my use of the apparently homogenous 'we' in these reflections.

This is a theme that has, regrettably, dominated too much academic discussion of this issue since we started talking about the Anthropocene a few years ago. This is not a book about academic disputes nor is it intended to reach a primarily academic audience. But because some of these notions have trickled out of the ivory tower and seeped like a noxious gas into the general culture, it is worthwhile to pause a moment and confront them.

The complaint is easy to state. It is that in employing the term 'Anthropocene' (which means the 'human age'), we have effectively erased all the politically important differences among groups of humans. In particular, 'we' are claiming that 'we' are all equally to blame for the climate crisis. Jason Moore, perhaps the most vocal and persistent of the critics, accuses those who use the label 'Anthropocene' of resorting to notions of an "abstract humanity," an "undifferentiated whole" of people (2015, 170).

The reason this alleged error is so important, we are told, is that it covers over the fact that climate change has been caused by a particular economic system, capitalism, and that its main perpetrators are the capitalists. Moore therefore wants us to rename the new epoch the Capitalocene. This suggestion triggered an avalanche of similar recommendations, among them the Misanthropocene, the Gynocene, the Cthulucene, the Plantationocene, the Homogenocene, the Anthrobcene and more.

These recommendations are generally aimed at asking us to notice an aspect of the new age that the term 'Anthropocene' might be concealing. We should *not* ignore the fact that the climate crisis has reached its current pitch via the economic intermediary of capitalism. The term Misanthropocene has presumably been suggested in an effort to remind us that the coming times will, in large measure, be hostile to human interests. Advocates of Gynocene want us to recognize that the new epoch has been the product of a patriarchal culture. All of the others just mentioned invite us, helpfully, to cast our gaze more widely than we might be inclined to do when looking at the impacts of anthropogenic climate change. Even so, we should reject the proposal to replace 'Anthropocene' with any of these alternatives. Why?

As climate writer Ian Angus has shown, the main reason is that the scientists who invented the term 'Anthropocene' have *not* in fact supposed that all humans are equally to blame for climate change (2016, 224–230). These scientists are alive to the issue of differential blame among various human groups for climate change. For example, one of the most useful concepts we now have for understanding our relation to the Earth system is that of 'planetary boundaries.' Scientists have argued that there are nine ecological boundaries: stratospheric ozone depletion, loss of biodiversity, chemical pollution, climate change, ocean acidification, freshwater consumption, land system change, nitrogen and phosphorous flow to the biosphere and oceans, and atmospheric aerosol loading.

With respect to each of these boundaries it is possible to quantify how close we are to exceeding the capacity of the Earth system to handle our environmental disruptions. Here, however, I'm less interested in the data themselves than in what the scientists thinking of the Anthropocene in these terms believe about an allegedly homogenous humanity:

> The current levels of the boundary processes, and the transgressions of boundaries that have already occurred, are unevenly caused by different human

societies and different social groups. The wealth benefits that these transgres-
sions have brought are also unevenly distributed socially and geographically.
It is easy to foresee that uneven distribution of causation and benefits will
continue, and these differentials must surely be addressed for a Holocene-like
Earth System state to be successfully legitimated and maintained.

(Steffen et al., 2015, 739)

This admits *everything* critics of the term 'Anthropocene' are trying to get us to
see with their variously tortured neologisms. All those critics are saying is that,
as these scientists put it, the new epoch involves an "uneven distribution of causation and benefits," and this must be addressed in any workable or just political
response we adopt to the challenges presented to us.

In any case, I believe it's crucial to stick with 'Anthropocene' precisely *because* it
forces us to focus on the challenges of the species as a whole. No matter where we
live on this planet, what socio-economic group we belong to or which ethnic, religious or gender identity we adopt, we will have to deal with the effects our activities are having and will continue to have on the Earth system. Nobody will be able
to sidestep the issue altogether. No matter how well-fortified, gated enclaves will
not protect the super-wealthy from desperate hordes, if it comes to that. And look,
as the quote from the planetary boundaries' scientists illustrates, it's not as though
adopting this species-level focus precludes asking tough questions about blame distribution, fairness and equity, the coming refugee crisis, the connection between
rising ethnic nationalism and resource scarcity, class conflict or whatever.

Those questions are key aspects of any attempt to understand what we are up
against. In my experience people are very receptive to the idea that the best way
to describe all of this is with a term that stresses, without mindlessly celebrating,
the new centrality of our whole species in the Earth system. That's exactly what
'Anthropocene' does. Such terminological accuracy contrasts sharply with, say,
'Gynocene' and 'Cthulucene.' Most people generally appreciate both the gener-
ality of the threat and the pressures it will put on our most important values. The
Anthropocene it must remain.

I don't want to sound too dismissive. Those who have sought to change the
epoch's name should be applauded for having raised the political stakes surrounding the issues we face. The responses of the scientists to them have sometimes been unwarrantedly testy; largely, I think, because the challenges are often
packaged in barely digestible, jargon-saturated prose (fans of the Cthulucene, I'm
looking at you). It's worth emphasizing that the value of inclusivity is a political
value first and foremost. As I have been saying, it is the foundational value of the
democratic enterprise. When we make the sorts of hard choices that await us our
decisions are thus intrinsically political ones. So let's talk about the ultra-rich.

Decisions about adaptation will inevitably be bound up with existing power
differentials within as well as across societies. I just said that nobody will escape
the ravages of climate change altogether. But the rich are sure as hell going
to try. Already, some Silicon Valley billionaires are making plans to flee to

New Zealand—temperate, prosperous and, best of all, *remote*—when the climatic going gets tough (Carville, 2018). The world's plutocratic class will attempt to exploit the crises we face in a way that benefits them exclusively; they will seek to extend their hegemonic influence over the shape of our future. The task of coming to terms with profound value conflict is thus inseparably bound up with that of seeing to it that we have measures in place blocking the ability of the super-wealthy to take advantage of the rest of us.

There is now much more climate change activism than there was when I began writing and talking about this issue some 10 years ago. That's great. However, as philosophers Ernesto Laulau and Chantal Mouffe argue, "new social movements exist in multiple forms which may be shaped through hegemonic struggle to progressive or reactionary ends" (1985, 87). Backers of reactionary ends will win the day unless the policies expressing their preferences are defeated by those expressing the ends of progressives. The problem is that—revolutionary situations aside—reactionary political groups, by definition, have more power than those seeking fundamental change.

This brute fact must inform all our decisions about technological adaptation to the climate crisis. If we ignore it, or fail to address it robustly, the ultrarich will ignore the adaptive challenges that confront the rest of humanity, so long as they believe they are protected from the devastation (whether or not the belief is warranted by the facts). Building a sea wall around New York is probably vital, but it is going to cost around $120 billion (US Army Corps of Engineers, 2020). But in the face of the demand to use their wealth to aid the collective the rich will either flee (hello, New Zealand) or stay but look for ways of removing themselves from harm while passing the wall's cost onto everyone else. The likely result is that the thing will never get built, or will be built shoddily, thereby forcing many people to put up with recurring catastrophes and the deprivations they bring.

This is one adaptation measure in one locale. But because the example generalizes, beginning more or less now we are in store for increasingly pitched conflicts over adaptation priorities, conflicts that will be driven largely by economic power imbalances. This has to do with differing capacities to avoid harm. But we must also beware of those who can profit from devastation and will thus attempt to abet it, the disaster capitalists about whom Naomi Klein (2008) has warned us. Either way, in the age of climate crisis ruling groups will try to do what they always do in such times: reproduce the social conditions enabling *them* to flourish. Our politics must evolve in whatever way is required to prevent this. In the next chapter, I'll be more precise about what this means.

Conclusion

If I've done my job properly so far you will have been persuaded to think of climate change as a deep—indeed, identity-altering—crisis for humanity. Crisis itself I have characterized as a matter of existential disorientation. This is an advance on the usual way of talking about crisis because it compels us to focus on everything at once. The whole social imaginary is up for grabs now. As I claimed in Chapter 1, this is evoking a certain culture-wide sense of bewilderment. Taken correctly, I suggested, there's nothing wrong with that attitude for it is true that we cannot *fully* explain either how we arrived at such an impasse or what the future has in store for us. All we know for sure is that the crisis is comprehensive in scope.

In this chapter I've explained what this entails for the values underlying the democratic enterprise. Insofar as we think we have made significant moral progress it is imperative to understand the nature of this progress so that we know how to focus our moral and political work in the times ahead. The progress we've made so far, I have argued, has mainly to do with expanding the circle of moral considerability to encompass those groups of people, and not just people, whom we had previously considered beyond the pale.

But inclusivity requires material support. In materially tough times, it readily gives way to its sinister other, exclusivity. And we are indubitably going to face materially taxing times in the coming decades. We are therefore going to have to *fight* for inclusivity. I'm not talking only about defending this value vigorously in public debate. As important as that will be, we should also be prepared to fight for inclusivity in a much more literal way. Because we will regress if current power structures are allowed to persist, and almost nobody relinquishes power willingly, our politics are going to get quite brutal for the foreseeable future. I see no way around this intensification of social and political antagonism.

Now it is time to change tack. So much of the literature on the Anthropocene talks about a fundamental rupture or rift in history. I understand the point of such language, but I want to push back against it just a bit. I have mentioned the possibility that we will be faced with more and more tragic choices. One of the advantages of the tragic outlook is that it allows for what the political philosopher Alison McQueen calls a "cautious optimism" about our prospects. This worldview tends to be associated with a cyclical view of history, where the same kinds of struggles and challenges recur. That means we can learn from past travails and shape our responses to current challenges accordingly. McQueen cites the Italian philosopher Niccolo Machiavelli (1469-1527) in this regard. Writing in the midst of the profound social and religious upheavals of renaissance Florence, Machiavelli remarks prosaically that "all people everywhere and always have led similar histories" (2018, 100).

In the next part of the book, I take this insight to heart. I have isolated five key events that many considered at the time they were unfolding to be world-shaking. I'm going to focus on the responses of a few canonical philosophers to these upheavals. In each case, the response uncovers a key intellectual innovation, one *we* cannot afford to lose sight of as we seek to reorient ourselves in the topsy-turvy world we have made. We begin, as all philosophy should, with Plato.

References

Angus, I. (2016). *Facing the Anthropocene: Fossil Capitalism and the Crisis of the Earth System.* New York: Monthly Review Press.

Boysen, L.R., et al. (May, 2017). "The Limits to Global Warming Mitigation by Terrestrial Carbon Removal." *Earth's Future* 5(5), 463–474. Retrieved from: https://agu pubs.onlinelibrary.wiley.com/doi/full/10.1002/2016EF000469. Accessed January 30, 2020.

British Broadcasting Service (BBC). (November 2, 2019). *Dresden: The German City that Declared a 'Nazi Emergency'*. Retrieved from: www.bbc.com/news/world-europe-50266955. Accessed November 3, 2019.

British Petroleum (BP). (February 14, 2019). *BP Energy Outlook 2019*. Retrieved from: www.bp.com/en/global/corporate/news-and-insights/press-releases/bp-energy-out look-2019.html. Accessed April 3, 2019.

Buchanan, A., and Powell, R. (2018). *The Evolution of Progress: A Biocultural Theory.* Oxford: Oxford University Press.

Burke, M., et al. (July, 2018). "Higher Temperatures Increase Suicide Rates in the United States and Mexico." *Nature Climate Change* 8, 723–729. Retrieved from: www.nature. com/articles/s41558-018-0222-x. Accessed April 18, 2019.

Burroughs, W.J. (2005). *Climate Change in Prehistory: The End of the Reign of Chaos.* Cambridge: Cambridge University Press.

Carville, O. (September 5, 2018). "The Super-rich of Silicon Valley Have a Doomsday Escape Plan." *Bloomberg*. Retrieved from: www.bloomberg.com/features/2018-rich-new-zealand-doomsday-preppers/. Accessed January 30, 2020.

Dufresne, T. (2019). *The Democracy of Suffering: Life on the Edge of Catastrophe, Philosophy in the Anthropocene*. Montréal: McGill-Queens University Press.

Emory, S. (February 6, 2019). "Deep Inside the Deadly Avalanche that Climate Change Built." *Wired*. Retrieved from: www.wired.co.uk/article/swiss-alps-avalanches. Accessed April 18, 2019.

Flynn, M. (April 19, 2019). "Three of the World's Most Elite Mountain Climbers Presumed Dead after Avalanche in the Canadian Rockies." *Washington Post*. Retrieved from: www. washingtonpost.com/nation/2019/04/19/three-worlds-most-elite-mountain-climbers-presumed-dead-after-avalanche-canadian-rockies/. Accessed March 3, 2020.

Ganopolski, A., and Rahmstorf, S. (2001). "Rapid Changes of Glacial Climate Simulated in a Coupled Climate Model." *Nature* 409, 153–158 (figure 4a).

Hampe, M. (2015). *Tunguska, or the End of Nature*. Chicago: University of Chicago Press.

Heath, J. (2002). *The Efficient Society: Why Canada Is as Close to Utopia as It Gets*. Toronto: Penguin Canada.

Heidegger, M. (1993). "The Question Concerning Technology." In *Basic Writings*. San Francisco: Harper-Collins, 307–342.

Intergovernmental Panel on Climate Change (IPCC). (2014). *Fifth Assessment Report, Summary for Policy Makers*. 1.1. Retrieved from: https://ar5-syr.ipcc.ch/topic_summary.php. Accessed February 1, 2020.

International Energy Agency (IEA). (2019). *Global Energy and CO2 Status Report 2019*. Retrieved from: www.iea.org/geco/. Accessed April 3, 2019.

International Monetary Fund (IMF). (2019). *Global Fossil Fuel Subsidies Remain Large: An Update Based on Country-Level Estimates*. Retrieved from: www.imf.org/en/Publications/ WP/Issues/2019/05/02/Global-Fossil-Fuel-Subsidies-Remain-Large-An-Update-Based-on-Country-Level-Estimates-46509. Accessed August 11, 2019.

Klein, N. (2008). *Shock Doctrine: The Rise of Disaster Capitalism*. Toronto: Vintage Canada.

Laclau E., and Mouffe, C. (1985). *Hegemony and Socialist Strategy: Towards a Radical Democratic Politics*. London: Verso.

McKinnon, C. (2011). *Climate Change and Future Justice: Precaution, Compensation and Triage*. London: Routledge.

McQueen, A. (2018). *Political Realism in Apocalyptic Times*. Cambridge: Cambridge University Press.

Moore, J. (2015). *Capitalism in the Web of Life: Ecology and the Accumulation of Capital*. London: Verso.

Oxfam International. (January 20, 2020). *World's Billionaires Have More Wealth than 4.6 Billion People*. Retrieved from: www.oxfam.org/en/press-releases/worlds-billionaires-have-more-wealth-46-billion-people. Accessed February 3, 2020.

Parker, G. (2013). *Global Crisis: Climate Change and Catastrophe in the 17th-Century*. New Haven: Yale University Press.

Steffen, W., et al. (2015). "Planetary Boundaries: Guiding Human Development on a Changing Planet." *Science* 347, 736–747.

United Nations Framework Convention on Climate Change (UNFCCC). (November 26, 2019). *Cut Global Emissions 7.6% Every Year for Next Decade to Meet 1.5° Celsius Paris Target*. Retrieved from: https://unfccc.int/news/cut-global-emissions-by-76-percent-every-year-for-next-decade-to-meet-15degc-paris-target-un-report. Accessed January 30, 2020.

US Army Corps of Engineers. (2020). *NY and NJ Harbor and Tributaries Focus Area Feasibility Study*. Retrieved from: www.nan.usace.army.mil/Missions/Civil-Works/Projects-in-New-York/New-York-New-Jersey-Harbor-Tributaries-Focus-Area-Feasibility-Study/. Accessed March 13, 2020.

Watts, J. (October 8, 2018). *We Have 12 Years to Limit Climate Change Catastrophe, Warns UN*. Retrieved from: www.theguardian.com/environment/2018/oct/08/global-warming-must-not-exceed-15c-warns-landmark-un-report. Accessed April 3, 2019.

Williston, B. (2015). *The Anthropocene Project: Virtue in the Age of Climate Change*. Oxford: Oxford University Press.

PART 2

Five intellectual innovations

4

PLATO: EPISTOCRACY

The harrowing Ridley Scott war film *Black Hawk Down* begins with this epigraph: "Only the dead have seen the end of war" (Black Hawk Down, 2006). As it turns out, there's a mini-controversy surrounding the source of that quote. In the film Scott attributes it to Plato, but there's no evidence of it in Plato's works. It was certainly said by the early 20th-century American philosopher George Santayana and repeated by General Douglas MacArthur in an address to cadets at West Point in 1962. Apparently, it's something of a meme among American soldiers. I guess it can help buck you up when you're in the middle of battle by suggesting that what you are going through is inevitable, at least for the living.

Some have suggested that because soldiers like the saying so much, it must be a paean to war, a celebration of death and destruction. For that reason, so goes the argument, we should not attribute it to the pacific Plato. This is all wrong. Most obviously, it paints Plato as some kind of pacifist, and I have no idea where people got that idea. Plato's *Republic*, a book we're going to look at in some detail in this chapter, constructs an ideal political society one of whose key elements is a *warrior class*. More importantly, even if Plato didn't say that only the dead have seen the end of war, he jolly well could have. As we'll see, it's a perfect description of how a pensive person might articulate his angst in the kind of world Plato *did* inhabit.

From the safety of a world at relative peace and when you've never experienced war yourself it's hard not to sound trite talking about the many scars war brings, but I'm going to try anyway. I've never known war in any meaningful sense. After the recent death of my stepfather, a Second World War veteran, there is now nobody in my family who has such direct experience either. Nobody involved in war comes out of it unscathed, but war's manifold pains are obviously heightened for those on the losing side of any conflict.

For these people, it can feel like everything important is now gone: property, loved ones, the homeland itself. We react in different ways to devastation on this scale. Some sink into despair or turn violently on those who led them down this dark path. Others, the luckier ones, are able to stumble with bent heads through the ravaged world. Nearly everyone feels a deep sense of estrangement and disorientation at such times, as though the whole world has broken free of its moorings.

Philosophers, those of them who have seen fit to write directly about war or who produced their best work in the midst of war, try to *think* their way through the disorientation. This often takes the form of what is known as metaphysical investigation, the study of the fundamental structure of reality. The very concept of metaphysics sometimes confuses non-philosophers, but it's easy to appreciate the enduring allure of the questions the metaphysician grapples with. I mentioned a few of these questions in the Introduction. One of the questions metaphysicians ask is whether or not this world around us is, as it were, all there is. Maybe there's a hidden world, one accessible to pure thought or to the saved.

Entertaining a possibility like that can sound like pure escapism, and so it often is. But not invariably. Sometimes it's a way of coping with collective trauma. In this case metaphysics, even the otherworldly variety, can be understood as the response to a pervasive sense of homelessness, wandering, exile, disunity and general existential angst. It is an attempt to find meaning in a world that has lost its sense of purpose, its basic principles of organization, an effort to set humanity on a new path by saying something novel about how we fit into the larger cosmic whole. Metaphysics is a form of homecoming in response to the perceived destruction or dissolution of the collective home. The kind of crises that can provoke it come in many forms, and we are going to explore several in this and the following four chapters.

Because war is probably the most visceral kind of collective crisis, we'll start with it. As I've said, a lost war can make the world appear suddenly strange to the losers. I don't mean strange in this or that isolated aspect but *thoroughly* strange, strange in a way that forces people to re-think just about everything—from the role of the divinity in their lives to the laws that order their interactions as citizens to the company they invite for dinner.

Plato, the first systematic philosopher in the Western tradition, responds to war's disorder in a way that might at first strike us as odd: by arguing that the knowers—which in his case means the philosophers—should be in charge of remaking and maintaining the social order. This is epistocracy (literally, the 'rule of the knowers'). My job in this chapter is to convince you that in the age of climate crisis our democracy must be epistocratic.

A generational clash

Plato's most famous dialogue, *The Republic*, written sometime around 375 BCE, is a lengthy discussion about justice. It is typical of Plato's dialogues. In almost all of them his philosophical mentor, Socrates, is presented as confronting conventional views about such topics as piety, courage, immortality, love and justice. Usually, an interlocutor will put forward a definition of the key term, which Socrates will then go on to demolish more or less mercilessly. At its best it is thrilling intellectual blood-sport.

One of the reasons for Plato's perennial appeal is that he is such a great writer, possessing an inimitable ability to convey deep philosophical truths through myth, metaphor, simile and allegory. I want to begin with a somewhat unorthodox interpretation of Plato's most famous myth, that of the cave. This myth—or simile, or allegory: it's really all of these things—is a uniquely arresting representation of Plato's whole metaphysical, political and moral vision. As such, it is, he says, a picture of "the enlightenment or ignorance of our human condition" (1955, 513–514).

Imagine that the world you inhabit, the one containing all the objects and people with whom you deal on a daily basis, is in fact a play. You are seated on a long bench, and have been for as long as you can remember, chained to a line of other people on either side of you. On a conveyor belt behind you runs a steady procession of mundane objects that are manipulated in various ways by those— the puppet-masters—in control of the procession. Behind those objects is a fire, which casts shadows of the objects onto a screen in front of you. You have no way to turn your head around and no reason to believe that what you see on the screen is anything other than the real world. Yes, Plato invented TV.

In this state, you are systematically deceived about your world. That is the human condition. The only hope is that one of these prisoners might turn her head, break free of her fetters, make for the opening of the cave and see what lies beyond. Most of us would resist this, so comfortable have we become with the illusions that define our lives. Here's what would happen if one of us were forced to look at the light of the fire behind us:

> And if he were made to look directly at the light of the fire, it would hurt his eyes and he would turn back and retreat to the things which he could see properly, which he would think really clearer than the things being shown him.
>
> *(Plato, 1955, 515e)*

Now, imagine the prisoners in the cave at a time in their collective life when the values and practices defining and holding together that life—the group's social imaginary—are for one reason or another beginning to fray, starting to seem less obvious to some citizens. There is burgeoning strife and folks are assembling warily into mutually suspicious or hostile factions.

One of the ways this might happen in any social order is along generational lines. So let's think of the cave in a state of rising political agitation and tension among its denizens, where one of the main social fissures is between the old and the young. Clarifying the nature of this clash can tell us a lot about the moral demands of our time, and about the very nature of morality itself. The opening scene of *The Republic* is immensely illuminating on this theme. In what remains of this section I will focus on it, then swing back to our own generational fissure.

The dialogue's participants have just returned from an Athenian festival at the Piraeus in honor of Bendis, the Thracian goddess of the moon. Two of them, Socrates and Plato's brother Glaucon, are assailed along the dusty path by the slave of Cephalus, an old and wealthy aristocrat, at whose house they are asked to gather. There, the discussion about justice unfolds among a number of Athens' best and brightest young men. As befits the veneration in which he is generally held in this group and in Athens generally, Cephalus offers the first definition. Justice, he says, is a matter of telling the truth and returning what we have borrowed.

Getting at the essence of a concept by attempting to define it is a very efficient procedure. This is because the definition is meant to apply to all instances or applications of the concept, and so a single counterexample refutes the definition. Someone might define swans as white waterbirds of the Cygnus genus. Sounds great, except for the inconvenient fact that there are black swans, counterexamples to the definition. Better try again.

Unfortunately for Cephalus, it's pretty easy to find counterexamples to his definition of justice. If I lie to my emotionally frail grandmother about her new hat, telling her how lovely it is when in fact looks like a baby squid resting on her head, most people will surely forgive me. They might even praise me for my discretion and diplomacy. As for returning what we have borrowed, Socrates has Cephalus notice that it would be highly imprudent to return a borrowed sword to its owner if that person had in the meantime become dangerously insane.

There you have it: two counterexamples, one for each part of Cephalus' claim about the essence of justice. Why do I say this exchange with Cephalus is highly significant? The reason has very little to do with his doomed definition. Rather, he symbolizes the old order, one that the young people are trying to find reasons to replace. After his definition is refuted he leaves the scene graciously, telling the others that he needs to attend to the particulars of an ancient religious ritual (a sacrifice). Cephalus's quick departure tells us that the prevailing ways of understanding right and wrong, and the specifically religious underpinning of many conventions and mores, will no longer fly.

So whatever else it is the battle for the soul of the city is generational. Of course, generations, the young and the old, are always battling for control of the future. The young typically feel as though the old are clogging things up, working into their dotage and thus taking all the good jobs, unable to see the value in new ways of talking, dressing, making music, etc. This sort of confrontation can sometimes provide just the tonic a complacent social order needs.

As I have said, Cephalus is both rich and old, and probably because of these two qualities he sees his duty chiefly in terms of the quest for existential ease. Here, for instance, is what he has to say about the value of money:

Now it is chiefly for this that I think wealth is valuable, not perhaps to everyone but to good and sensible men. For wealth contributes very greatly to one's ability to avoid both unintentional cheating or lying and the fear that one has left some sacrifice to God unmade or some debt to man unpaid before one dies.

(Plato, 1955, 330–331)

Columbus said something similarly crass, observing in his journal that "gold is most excellent because he who possesses it can do as he wishes in the world. It can even drive souls into paradise" (quoted in Thiele, 2019, 108). Money is a form of power the possession of which enhances one's ability to act rightly.

Do you feel as though you have wronged a colleague at some point in the past, maybe by lying about that colleague's performance so that you could curry favor with the boss? Just buy the guy a boat and all will be forgiven. Think you might have fallen short in your outward displays of religious devotion over the years, perhaps because you were too busy spending all your dough on yourself? An extra fat lamb at this year's spring sacrifice should appease the divinities. And so on.

One way of responding to a view like this is that of Socrates himself. It is to deny the connection between wealth-as-power and morality. The alleged connection does look pretty dubious. In examples like the ones just canvassed most of us believe that genuine moral debts cannot be cancelled through acts of strategic financial largesse on the part of wrongdoers. This is true, we might think, even if the wronged agent is perfectly happy with this outcome. For one thing, assuming the wrongdoer's cash doesn't run out, the proposed resolution does nothing to prevent his future wrongdoing. Although this describes the 'morality' of many corporations pretty well, most of us expect more from other people.

In any case, Socrates' interlocutors in *The Republic* have a more interesting response than this to Cephalus. They don't begrudge the old man his wealth, nor—more importantly—do they deny the connection between wealth-as-power and morality. Rather, they dismiss him for being so unambitious. A bit earlier, for instance, Cephalus had been talking about how happy he is to be old since the old are no longer troubled by the demands of sexual desire. Lacking desire's anarchic thrust, men like Cephalus will never have to worry overmuch about paying amends for wrongdoing because they likely won't wrong anyone very gravely.

Cephalus is treated with respect by the other conversational players here, but running beneath this respect is a just detectable current of contempt for the tiredness of this view. In contrast to Cephalus these boys—especially the ferocious Thrasymachus, the inventor of the 'might makes right' doctrine—have fire in their bellies. But precisely because of this, and also because they endorse his claims about the connection between power and morality, they are supremely dangerous, whereas Cephalus is anything but. To one degree or another, they are all impressed by the regime in Sparta and are wondering out loud whether or not that city might be a political model for Athens. Some of them long for the unsullied happiness enjoyed, they surmise, by tyrants. This conversation is thus no abstract exercise for these men. As members of the political elite, the establishment of any sort of undemocratic regime could thrust *them* into positions of virtually limitless power.

As Socrates points out, Cephalus's definition of justice is far too narrow. It is, we might say, overly juridical. At bottom, he expresses what many people take to be the requirements of the law, a law written by and for those already in power. There's nothing intrinsically wrong with obeying the law and following custom, of course, but notice what this move assumes. It takes for granted that the social and political order, and therefore also the legal order, are on solid cultural ground.

I find this part of Plato's dialogue so engrossing because it encapsulates deep truths about the way folks fight for philosophical terrain in times of social upheaval. The climate crisis has now become a generational battle over the meaning of justice. Greta Thunberg and her allies are challenging an older generation that appears too content with the way things are. It's an entire social stratum composed of characters like Cephalus. In a word, it's the powerful. Even if such people admit that climate change is a problem, too many of them are incrementalists in their approach to it. Implicitly or explicitly, they reject the idea that fundamental, systemic change is required.

More particularly, they believe that endless economic growth will ultimately save us from whatever problems it itself has engendered. Greta *et al.* reject precisely that article of faith. Think of it as analogous to Cephalus's belief about the nature of justice and you will see what makes this generational cohort similar to Socrates. Both know that the old order is collapsing and they are trying to get everyone to think correctly about justice in light of this reality.

That's what's transpiring right now in the cave, *our* cave. Plato can help us appreciate two key features of the situation so construed. First, that the nub of our generational clash has to do with the connection between morality and power. Second, that because so much is at stake the clash is both unavoidable and perilous. Even as we reject the business as usual of Cephalus we must beware of all-too-real political characters like Steve Bannon, Trump's oily former *éminence grise*, channeling Thrasymachus from the culture's edge. Waiting to pounce.

We aren't done with the Greeks yet. To appreciate fully their perplexities about justice we need some wider stage-setting than *The Republic* itself gives us.

Weaned on war

My parents were pretty apolitical. They resolutely put into practice a belief that seems incomprehensible to me now but was apparently standard in the average 1960s and 1970s middle class North American suburban home: don't talk about politics or religion at the dinner table. As a result, I don't remember ever hearing about, say, Vietnam while I was growing up. My own house, I'm happy to say, is not like this. We have regular battles about politics and religion over our fish and chips. These exchanges can get heated, but it's worth it. My kids—13 and 17 at the time of writing—know infinitely more about these subjects than I did at their age.

They even know something about the war that has always interested me the most: the Peloponnesian War, a conflagration that shaped Plato's world from top to bottom. Plato was born in 427 BCE, at a time when his city-state was four years into a military conflict with Sparta that lasted from 431 BCE to 404 BCE. He was weaned on war. But not just any war. This was a vicious, nearly-30-year contest between two foes, and their allies, with the biggest military forces the world had ever seen. The Peloponnesian War, and its aftermath, engulfed Athenian society for a good chunk of Plato's life.

Coming back to Ridley Scott's epigraph, the sheer persistence of this war might make anyone enduring it believe that this state of affairs is the lot of living, and that only death can bring relief from it. If you doubt this, just broaden your definition of war. There's no reason to confine the state of war within the frame of its formal declaration and cessation, let alone to the actual fighting. The 17th-century English philosopher Thomas Hobbes (1588-1679) said that a state of war is like lousy weather. It's the constant *threat* of rain and wind that calls forth the description of foulness. The same goes for war. Its anxieties and challenges reverberate well beyond the tidy dates that get inscribed in the history books.

It is revealing that *The Republic* is written some 30 years after the official end of the war. Plato and his friends are still grappling with its consequences, including the possible breakout of more explicit violence. It's not hard to see why. This war brought Athens from a position of world domination and cultural supremacy to fragmentation and humiliation. It is crucial to appreciate just how dominant the Athenians were at the beginning of the war, and how thoroughly this position had, as we say, gone to their heads. They were a supremely self-satisfied people. They believed that their power was evidence of their cultural superiority, and they expected every other city to bow before them.

There is no better encomium to the Athenians than the famous funeral oration of its leader, Pericles, given in the winter of 430 BCE. The funeral was held as a tribute to "the first men to die in the war." Here's a snippet of it:

> Athens alone of cities today outdoes her reputation when put to the test. . . . We shall be the wonder of this and of future generations. We need no Homer to sing our praises, nor any poet to gratify us for the moment with lines that may fail the test of history, for we have forced every land and sea to yield to our daring and we have established everywhere lasting memorials of our power for good and ill.
>
> (Thucydides, 2013, 113–114)

By the way, that bit at the end about doing ill does not mean that the Athenians regretted anything they did. It is not bad from the point of view of the Athenians.

It is instead an expression of the common view at the time—one that comes up for quite a bit of discussion in *The Republic*—that it is right to bring ill to one's enemies. The 'ill' is what *they*, your enemies, don't want to see happen to them, like having their navy destroyed, their city burned and their citizens slaughtered. What's bad for them is thus good for you. Adopting the correct perspective, it turns out, is all-important in understanding right and wrong.

When the war started, Athens was in the midst of its Classical Age, a period that saw the creation of some of the West's greatest cultural achievements in art, philosophy, architecture, politics and theatre. Because of its naval prowess, it had become a commercial powerhouse, establishing trade routes from Britain to India. Politically, it was at the center of a huge empire or confederation, led initially by Pericles, until his death in 429 BCE (just a year or so after he delivered the funeral oration).

By the end, the empire had collapsed, the democracy undermined by a series of self-serving oligarchies and tyrannies. The shining imperial jewel of Greek culture and commercial power had become a subject city in a political space dominated by its foes. Eventually, something like the old democratic order was restored in Athens but it stood on extremely shaky ground. Tyranny and general political extremism lurked menacingly in the cultural background as a perennial possibility.

In the midst of this uncertainty and anxiety, in 399 BCE, the Athenians executed Socrates, who had been charged with impiety and corrupting the youth. Was he guilty? In some sense, he probably was. He would have no truck with those who think we can look to the gods or conventional pieties to tell us how to behave with honor. Claims of this importance must be backed by reasons. And just because of this stance, which he did indeed press upon the young (among others), Socrates was a royal pain in the ass to the establishment. Indeed, he could with some justification be seen as a threat to the ideal of democratic moderation which some Athenian worthies were seeking to keep alive in the face of a constant threat emanating from Sparta.

In his masterful history of the Peloponnesian War, the ancient Greek historian Thucydides reports on the civil strife which had already engulfed so many city-states as early as 427 BCE. "Practically the whole Greek world," he tells us, "was in turmoil" at this time. The turmoil took the form of rival political gangs seeking to impose their will on the rest of the city, and thus further their narrow interests at the expense of the whole. Thucydides goes on:

> The leaders in the various cities would each of them adopt specious slogans professing the cause either of 'political equality for the masses' or 'aristocracy-the government of moderation;' they pretended in their speeches to be competing for the public good, but in fact in their struggle to dominate each other by any available means they brazenly committed all manner of atrocities . . . with no regard for the constraints of justice and the public interest.
>
> *(2013, 212–213)*

Again, that's Thucydides' diagnosis of what is going on virtually everywhere at the time. In the smaller cities, these struggles had mainly to do with various factions trying to curry favor with or build up the power of either pro-Spartan or pro-Athenian elements within the city. But the rot had spread even to Athens itself.

In his *Seventh Letter*, purportedly written when he was very old (though the authenticity of this document is in some doubt among scholars), Plato gives voice to much the same complaint about his own city:

> When I considered all this, the more closely I studied the politicians and the laws and customs of the day . . . the more difficult it seemed to me to govern rightly. Nothing could be done without trustworthy friends and supporters; and these were not easy to come by in an age which had abandoned its traditional moral code but found it impossibly difficult to create a new one. At the same time law and morality were deteriorating at an alarming rate, with the result that, although I had been full of eagerness for a political career, the sight of all this chaos made me giddy.
>
> *(1955, 16)*

Early in *The Republic*, the new way of thinking is expressed succinctly by that arch political realist, Thrasymachus. Socrates labors assiduously to undermine the view that might makes right, but in the end cannot refute it head-on. He abandons the discussion with Thrasymachus and, throughout the rest of the book, tries instead to show us directly what a better world might look like. But Thrasymachus's shadow lurks over the whole enterprise, and indeed over our own attempts to understand the power-morality nexus.

Why are we so reluctant to admit that our civilization is headed full speed toward the ecological cliffs? Plato's answer is that the shadow play of our civilization—for instance in the form of consumer baubles dancing before our eyes—is being stage-managed by people who have taken Thrasymachus's message to heart. Because they believe there's no gap between power and morality, and they have the power, they see no problem in perpetuating a system that is designed primarily to serve their interests. For Plato, there is only one solution to a political state of affairs that has degenerated this completely: to make government answerable to the political whole.

The *Seventh Letter* puts the problem succinctly:

> I was forced, in fact, to the belief that the only hope of finding justice for society or for the individual lay in true philosophy, and that mankind will have no respite from trouble until either real philosophers gain political power or politicians become by some miracle true philosophers.
>
> *(Plato, 1955, 16)*

What sort of society are the philosophers meant to concoct? Plato's ideal republic is based on the notion that everyone should do their part, should stick to that function to which they are suited by nature. This becomes the very definition of justice for him.

There are three classes: the philosopher-kings, the guardians or auxiliaries, and the artisans. The first group governs, the second maintains internal and external order, and the third makes the economy chug efficiently along. Rather than dismissing this construct as outrageously unfitted to our times, we might focus on two aspects of it that *are* relevant for us. The first is that there is no significant economic inequality in it. Indeed, it's the quest for disproportionate wealth by one socio-economic faction that corrupts every kind of political order. Plato will have none of it in his ideal republic. Second, the structure is maintained by those who have an objective grasp of what is good for the whole. The latter point makes Plato an epistocrat, someone who believes that the knowers should also be the rulers. Time to investigate this crucial concept.

Epistocracy

Come back to the cave for a moment. After describing how it is that a person might turn away from the play of false images—perhaps the most difficult moment of enlightenment, the very first step—Plato goes on to say that if she could somehow get beyond even the fire, and see the light outside, the sight would surely stagger her. Still, if she perseveres, she might be led even further, to the whole realm outside of the cave, then eventually "the heavenly bodies and the sky itself at night." The last step, at least as far as *this* world is concerned, would be to look up to the sun, "and gaze at it without using any reflections in the water or in any other medium, but as it is in itself" (1955, 319).

What is Plato telling us here? Not that we should stare foolishly at the actual sun, as though in doing so we could somehow reach the pinnacle of wisdom rather than (or perhaps in addition to) frying our eyeballs. This is a simile, and the 'sun' in it represents the Form of the Good, goodness itself, the paradigm of all the particular good things. It is the unchanging essence of goodness in which such things—a good person, a good city, a good meal, a good horse, a good bath—must 'participate' to earn the description, 'this is good.'

For Plato, every concept has an eternal and unchanging Form like this. The idea applies not just to Goodness, but also to Equality, Beauty, Big-ness, Horse-ness and Charioteer-ness. These Forms exist beyond the material world, in a heavenly realm to which only the mind of the properly trained philosopher has access. The discipline of philosophy, wedded to pure mathematics, is supposed to catapult us to a self-sufficient realm beyond all materiality so that we might bathe in the pure essence of all things. Pretty heady stuff.

There are numerous problems with this move, many of which were anticipated by Plato himself (another mark of his brilliance). It would take us too far afield to consider any of them here, so in what follows we will focus instead on what the philosophical ascent to the Forms entails for our understanding of knowledge and truth. For Plato, the highest Form of all is that of the Good.

Scholars are notoriously divided about what this might mean, so anything we say about it will be provisional at best. But here's a thumbnail interpretation.

To say that the Good is the highest Form means that it encompasses all the others. It can be helpful to think of this as analogous to the way some theologians and philosophers have thought about God. God, on this conception, is the Being whose Knowledge contains all lesser items of knowledge. God sees how all of the elements of the whole are connected. You don't need an anthropomorphized deity to make sense of this idea. The Platonic Good is simply that perspective from which one is able to see how absolutely everything hangs together.

Fortunately, we needn't endorse the idea that there is a Form of the Good to appreciate what Plato is getting at. To reprise a set of terms from Chapter 2, he is telling us that any sentence we formulate about the world has truth conditions whose content is not up to *us* to decide. This makes him an *objectivist* about truth. Because it is so important, let's linger briefly on this.

Contemporary philosophers have articulated Plato's point by asking about the appropriate 'direction of fit' between our desires or beliefs on the one hand and the world on the other (Searle, 2003). With desires, the direction of fit is person-to-world because desire seeks to alter the world to meet its demands. If I desire a drink, then I will want the world transformed from, 'Excuse me, but I have no bloody drink in my hand!' to, 'I now have a drink in my hand, thank you very much.' The world, thank Christ, has been made to fit my desire.

With beliefs, the appropriate direction of fit is exactly the opposite, world-to-person. If there is in fact no drink in my hand, I may curse my maker for this absurd state of affairs, but I should not believe that there *is* a drink in my hand. Our beliefs should fit the world. Incidentally, we really need to clean up the way we talk about belief. No doubt wanting to distance ourselves from what the religious say about it, we've come to understand belief as an intrinsically irrational thing. But it's not. Belief is simply the psychological attitude of assent to a proposition. The person who says that the earth is 4.5 billion years old is expressing a belief no less than the person who says it is 6,000 years old. The difference between the two is that the first belief is rational (because it's evidence-based), while the second is not (because it's not).

Even if you are loathe to allow professional philosophers to call the shots—and I wouldn't particularly blame you for this—you should nevertheless assent to the more fundamental claim Plato is making here, namely that *ruling must be constrained by knowing*. That's the essence of an epistocracy: rule by or through the truth-seekers, those whose job it is to grasp the whole. In this respect we surely need more of Plato if we are to confront the profound challenge to truth that defines the climate crisis. Just ask the climate scientist Michael Mann.

One day in August, 2010, Mann was opening his mail and out of one envelope spilled a white powder, which turned out to be corn starch but was clearly intended to make Mann believe it was anthrax. The resort to a (fake) biological weapon against Mann was the culminating event of nearly 20 years of harassment from people who evidently think his simple and indisputable message—the planet is warming dangerously—must be suppressed by any means necessary.

Nor is the harassment of Mann an isolated event. Around the world, but especially in the US, climate scientists are under attack from those unhinged members of the political establishment, as well as their deranged supporters, who believe that climate science is a threat to the American way of life, real Christianity, etc. With Trump, the problem has gotten much worse:

> Past administrations and their allies have falsified, fabricated, or suppressed evidence . . . let political considerations drive science advisory board appointments, targeted essential data collection initiatives for elimination, and intimidated, censored, and coerced scientists. Under the Trump administration, these threats to the federal scientific enterprise have escalated markedly.
>
> *(Center for Science and Democracy, 2017)*

The Trump administration's woefully inadequate response to COVID-19—including dissolving the Global Health Security and Biodefense Unit in 2018—underlines the accuracy of this claim. It may be that the awful toll of the pandemic, demonstrably worsened by deliberate misinformation coming right from the top, will persuade Americans to arrest this self-destructive epistemic trend. But I'm not holding my breath. In any case, it is exactly what we should expect when rulers are not properly constrained by knowers.

Plato was aware of and deeply worried about politically motivated distortion or suppression of truth. Throughout his career he directs sustained philosophical fire at the so-called sophists. These were teachers of rhetoric who would hire themselves out to those wanting to learn how to use language to gain power and wealth. There are numerous critiques of the sophistic enterprise in the dialogues. The most persistent one is that the sophists effectively reduce truth to power. In a city trying to reconstruct its institutions and sense of moral purpose in the wake of war, Plato believed this to be a disastrous trend. In fact, it is not too much of a stretch to say that it is war by other means. For what is war if not the will to power's reduction of every element of collective life to its blunt demands?

The message is every bit as vital for us as it was for Plato. We are undoubtedly in the midst of a crisis of knowledge that is unprecedented in the modern democratic age. It is amazing that Donald Trump can lie as brazenly and as often as he does with no significant hit to his popularity. Those who are unmoved by this behavior seem incapable of making a distinction between truth and falsity. Whatever their leader says must be true. In 2018, Bannon proclaimed that the Democrats should not be the chief concern of Republican strategists. "The real opposition," he said, "is the media and the way to deal with them is to flood the zone with shit" (Illing, 2020).

Flood the zone with shit. In a world that has adopted this epistemic standard there is no daylight between truth and power. Whoever has a sufficient quantum of raw power—political, economic, military—can easily point to 'corroboration' of their worldview in the muddied swirl of manufactured facts and data. That's exactly what Plato might have said about the sophists. More to the point, it's exactly what organized climate change deniers sought to do when they asserted that their "product" is "doubt" (Oreskes and Conway, 2011). Not falsity per se, just the sort of deep and pervasive confusion that gives license to political and economic business as usual.

We should therefore not be too resistant to Plato's central message: that the best way for socially critical knowledge to thrive in society is to put those who possess it—or at least those functionaries who will reliably do their bidding—in charge. That's all there is to the idea of epistocracy. Moreover, putting the point this way allows us a degree of openness about who exactly the knowers are. Climate scientists, obviously, but also those working in the relevant parts of the humanities as well as indigenous peoples, and many others. Even so, there's a limit to this openness. One thing that jumps out about nearly all climate deniers is that their political identity determines their stance on the science. Sometimes, all you need to do is look at the social media profile. The political allegiance—often some form of libertarianism—is front and center. These people should be actively excluded from shaping the prerogatives of government.

Of course, it's easy to see what has gone epistemically wrong with these folks: they've got the direction of fit backwards. Because dealing with climate change will, they suppose, lead to socialism, and socialism is undesirable, the science must be cooked-up. If you start with this bit of tortured logic, you can spin out all the derivative fantasies in which deniers typically indulge. One of the funniest, but also most telling, of these fantasies is the idea that climate scientists are pulling the wool over our eyes because they are so eager to tap into those heaps of grant money lying all over the place. If you've ever met a scientist, you'll understand why I call this view funny. It is telling because of the way it projects the denier's view of the relation between truth and power onto everyone else.

That view is fundamentally ideological. 'Ideology' is a contested term in the humanities and social sciences, and I don't want to get bogged down here in the mostly stale controversy over its meaning. But one aspect of ideology shared by nearly all interpreters of it has to do with the attempt, by those in positions of power, to make a partial view look like a view of the whole. That is, ideology is a tool used to advance the interests of a segment of society in a way that *falsely* portrays those interests as universal. Think of trickle-down economics.

If ideology is all there is in political life, then the denier is right: the scientists are trying to dupe the rest of us. But there is no reason to believe that ideology is inescapable. In fact this belief is itself an expression of ideological thinking. Flooding the zone with shit is an ideologically motivated attempt to get people believing that there's nothing but ideology. At least in the case of climate change denial, this cynical maneuver has now been thoroughly exposed for what it is. Plato offers an alternative to it, one we cannot afford to ignore.

The fate of democracy

The vision of the Forms is, we are told, so splendid that those who partake of it will want to remain in this ethereal realm forever, lost blissfully in the uninterrupted contemplation of eternal truth. But Plato, through Socrates, argues that these potential truth-addicts must be ripped away from contemplation and forced to come back to the cave in order to liberate their former fellow-prisoners. The lawmakers in the ideal state will point out to these would-be intellectual layabouts that they have been bred both for their own good and for that of the larger community, and that because they "have seen the truth about things admirable and just and good" they must return to the cave and attend to the thankless job of building a republic (Plato, 1955, 342).

Plato knows that the purely moral exhortation is unlikely to work, however, so the lawmakers also point out that if the philosophers refuse to rule, people worse than them will. And one of the things these people might do is muzzle, perhaps even hunt down, the intellectuals. Is this mere hyperbole? You might think so if you understand 'philosophers' in the narrow sense, meaning professional philosophers. These days, most professional philosophers are entirely unthreatening to the establishment. But it is less obviously hyperbolic if you think of 'philosophy' in the broader sense I have been urging on you in this chapter, where the philosophers are the knowers, and where the knowledge they have threatens vested economic and political interests.

Even so, some people are tempted to abandon Plato at precisely this point. The philosopher Karl Popper devotes an entire volume of his massive work *The Open Society and Its Enemies* (2013) to lambasting Plato on this score. The book was published as the world was emerging from the Second World War, just waking up to the full scope of the horrors perpetrated by the Nazis. Popper's analysis is still very influential, so it's worthwhile to linger on it for a while.

We cannot dismiss altogether this sort of reaction to an ancient philosopher who was unwaveringly anti-democratic. How could one argue any other way in 1945? But we must also remember that Plato was as opposed to tyranny as he was to democracy. In fact, he thought that tyranny was much worse than democracy. The problem with democracy, he believed, is that it tends to produce tyranny by facilitating the rise of demagogues.

Why? We are accustomed to think of democracy as rule by the people, and consider this an unambiguously good thing. But think of the other side of this same coin. Because democracy encourages the belief among everyone that their opinions and desires matter, they will often seek out someone to represent those views, to give them real material expression in the political world. Again, there's nothing wrong with this in principle, except that democracy itself does not discriminate morally among these opinions and desires. For example, if a group of people is consumed by resentment because times are hard, and they manage to fix on an ethnically identifiable scapegoat for their hostility, enterprising demagogues will seize the opportunity to give voice to these negative emotions.

Finally, once they obtain power demagogues tend to want to consolidate it by any means necessary, i.e., to establish a political tyranny. The point is empirically verifiable. As Alexander Hamilton put it in the *Federalist Papers*, "history will teach us that of those men who have overturned the liberties of republics, the great number have begun their career by paying an obsequious court to the people; commencing demagogues and ending tyrants" (Hamilton, 2003, 29). Now come back to Popper. He is consumed by a single question: what in our shared culture could have allowed for the Nazi rise to power in Germany? And then he draws a more or less straight line from Plato to those events.

But let's look a little more carefully at some of the events. As we know, the Nazis were able to consolidate their hold on German politics through legally ordinary, democratically above-board means. This was a conscious decision on Hitler's part after his party tried to seize power through a coup d'état in 1923, the infamous Beer Hall Putsch. The attempted revolutionary seizure of power failed and Hitler was thrown into jail for nine months, during which time he composed *Mein Kampf*. He did not give up his dream of running the show all by himself, however. He simply decided to obtain it by working the levers of the system, which he did indefatigably for the next 10 years.

Eventually, in January of 1933, he was appointed Chancellor by Paul von Hindenburg who had tried, unsuccessfully, to confer the office on several others before coming round to Hitler. The Enabling Act, which gave Hitler the power to enact laws that bypassed the Reichstag, was passed by the cabinet just two months later. This effectively established Hitler as the perpetual dictator of Germany. For the previous 20 years this arch-demagogue, enabled by democratically endorsed liberties, had been whipping up the support of violent anti-Semitic mobs. Thus the stage was set for the most destructive political tyranny the world has ever seen. And none of it could have happened without the helping hand of ordinary democratic politics.

To repeat: Plato says that democracy enables demagoguery and demagoguery leads to tyranny. Now, which perspective provides the more astute interpretation of events like those that unfolded in Germany in these years, Plato's or Popper's? Alright, but surely Popper has a point? Even if Plato has a pretty good grasp of the way democracies can degenerate into tyrannies, it doesn't follow—does it?— that we should embrace his rejection of democracy. The answer to this question is more complex than it might seem, part of the complexity having to do with the different ways we now understand some of the key political terms at issue.

We have seen that Plato's friends in *The Republic* admire the Spartan rulers. Sparta was a mixed regime, combining elements of oligarchy and timocracy. Timocracy is rule by a warrior class, one governed above all by considerations of honor and glory and the pursuit of military dominance. Next to the epistocratic aristocracy Plato preferred, this was the best sort of regime. But it too tended to degenerate, in this case into oligarchy, the rule of those seeking solely to expand their personal wealth. In a sense, Sparta was caught in the middle of this transition.

This is the pattern we find throughout Plato's political philosophy: given the foibles of human nature, all political regimes, including epistocracy, have a tendency toward instability and degeneration. This is why, although he is often treated as a starry-eyed utopian thinker, Plato is in fact a realist. Or better, his idealism is always tempered by his realism about what is politically possible.

For one thing, he had first-hand experience of political extremism after spending some time later in his life at the court of the tyrant Dionysius I of Syracuse. Closer to home, he never forgave the Athenians for what they did to Socrates and this deeply affected his ability to trust any unschooled political class. The allegory of the cave is a (very) thinly disguised story about Socrates. Most people would respond to the philosopher who ascended the heights, then came back down again, by saying that "his visit to the upper world has ruined his sight." Surely they would "kill him if they could lay hands on him."

And so they did. Plato is ultimately skeptical about the chances of success for his ideal republic. Rather than a real political possibility it is, he concludes, probably best considered an ideal "laid up as a pattern in heaven where he who wishes can see it and found it in his own heart." In other words, Plato was highly attuned to the political realities of his day. Judgements of better and worse, applied to existing political regimes, were thus for him at least partly relative. Of course, they all fail when compared to the ideal, but saying this does not preclude also noticing that, say, the timocratic/oligarchic regime in Sparta is *better* than the tyranny in Syracuse. If we are going to employ Plato's insights to help us better understand our own predicament, we need to remember this.

Here then is the question to ask: what is now the biggest threat to global stability, justice and a sustainable future for both humanity and the non-human biosphere? The answer, I think, is clear. It is not democracy but oligarchy wedded to toxic ethno-nationalism. In their penetrating analysis of potential future political scenarios, *Climate Leviathan*, Geoff Mann and Joel Wainright identify a reactionary tendency in contemporary politics that they label "Climate Behemoth":

> In the United States, the signature affiliations of the reactionary right—market fetishism, cheap energy, white nationalism, firearms, evangelical faith—buttress reactionary Behemoth. The result is an opportunistic, but contradictory and unstable, blend of fundamentalisms: the security of the homeland, the freedom of the market, and the justice of God.
>
> *(2018, 46)*

Unlike these authors, I have no idea just how "unstable" this form of reactionary conservatism is, but I do think it is the key threat to democratic values. As climate catastrophes mount and more pressure is put on basic ecosystem services the global financial elite is clearly positioning itself to hoard ever more of the world's diminishing resources. There is no other way to understand the meaning, and the upshot, of skyrocketing inequality.

But a pointed ecosocialist consciousness is *also* emerging now. It is manifest in the rising tide of ecosocialist thinking among writers like Naomi Klein (2014) and Jedediah Britton-Purdy (2019), in policy-relevant documents like the Green New Deal, and in the global upsurge of climate activism among groups like the Sunrise Movement and Extinction Rebellion.

The contest between epistocratic ecosocialism and oligarchic reaction is the defining political antagonism of our times and there is no reason to believe that the ordinary, deliberative politics of liberal-democratic consensus can resolve it. At least in its post-WWII articulation, liberal-democratic philosophy has rested on the notion that disparate social and economic blocs could forge institutions that served everyone's interests. They wouldn't all necessarily be happy with the settlement, but at least they could enact a stable *modus vivendi*.

This optimism, according to Katrina Forrester, was rooted in the belief in "continued growth and lasting stability," a belief that is not "capable of fully making sense of the current conjuncture" (2019, 277). Although Forrester does not put the point this way, *the* flashpoint of the current conjuncture is the climate crisis. The battle lines of this crisis have therefore been drawn, even if the two sides are still in the process of coming fully to form. Because he was above all concerned with combatting political regimes characterized by (a) steep economic inequality and (b) indifference to knowledge of the whole, I know which of these sides Plato would have been on.

Conclusion

Plato's novel answer to the malaise of his times is the extraordinary suggestion that unity can only come from an even more fundamental fracture. Only by showing that at the most basic metaphysical level reality is comprised of two realms—the world of Forms and the ordinary world down here—can humanity step back from the abyss of total meaninglessness.

This is metaphysical *dualism*, the idea that reality is split into two basic kinds of things or places. Plato will ultimately say that one of them is fully real while the other is not: Being and non-Being, the Sun and the shadow-play. This is, perhaps, the ultimate paradox of Plato and all who follow in his intellectual path: unity through fundamental bifurcation. Metaphysical dualism thus answers the question of how we fit into the whole, even if only a select few get a glimpse of this whole's structure.

Plato is responding to a crisis at the level of the whole world, or at least what *seemed* like the whole world to war-weary Athenians. In a sense, this seeming is all that really matters. In the final period of their disastrous collective life, the Easter Islanders—never more than 20,000 souls—must have felt the same way even though there was a thriving world beyond theirs. They cut down all the trees on their tiny island to help them build statues that would appease their gods, then died because they couldn't make new fishing boats when the old ones rotted.

It is often asked what was going through the heads of those who cut down the very last tree. Who knows, but long before that sad moment they must all have been feeling profoundly disoriented. It's this sense of disorientation I want to highlight here. It will come up again and again in the chapters that follow. It's an existential crisis, which is why philosophers and religious leaders have always felt compelled to conceptualize it. The specifically philosophical response to such crises is to remake the shattered whole in thought, then hope this reconstruction works its way into the real world. In the process, these thinkers have supplied us with an invaluable idea or ideal. And I mean invaluable for us in the precise context of the climate crisis.

Plato in particular bequeaths us a view of the place of knowledge in the organization of society which we can no longer afford to ignore, if we ever really could. It is this: that if we are to survive in anything like the state to which we have become accustomed—or just avoid slipping into barbarism—our leaders must be constrained by knowledge of the ecological whole in which our political collectives are embedded. Our democracy must *also* be an epistocracy.

As valuable as this lesson is, there is nevertheless something missing in Plato. The stress on truth and knowledge is, we might think, too intellectual, *overly* cognitive. It's worth recalling that the study of mathematics is a central component of the education of Plato's philosopher-kings. We might therefore wonder how this very abstract training is capable of motivating those who receive it to *care* adequately about anything other than doing more math. Perhaps one of the reasons the lawmakers find it difficult to convince the philosophers to govern is that they, the lawmakers, have neglected, or misunderstood, this fundamental motivational component of the story we are telling.

How do we correct this? Ultimately, the best way to overcome it is to resist the dualism on which it is based. The very idea of constructing a world entirely apart, into which the aspiring knower must reach for her intellectual treasure, is, we might suppose, the root of the trouble. But we mustn't move too quickly. We can be dualists and still talk more pointedly than Plato does about what it might mean to care genuinely about the whole we are trying to understand and manage. This requires us to introduce a concept that is entirely missing from Plato's political philosophy: love. Some 800 years after Plato, a Neoplatonic Christian intellectual, destined for sainthood, figures this out.

References

Black Hawk Down. Film (2006) Directed by Ridley Scott. Culver City, CA: Sony Pictures Entertainment.

Britton-Purdy, J. (2019). *This Land Is Our Land: The Struggle for a New Commonwealth*. Princeton: Princeton University Press.

Center for Science and Democracy. (July, 2017). *Sidelining Science Since Day One*. Retrieved from: www.ucsusa.org/sites/default/files/attach/2017/07/sidelining-science-report-ucs-7-20-2017.pdf. Accessed April 12, 2019.

Forrester, K. (2019). *In the Shadow of Justice: Postwar Liberalism and the Remaking of Political Philosophy*. Oxford: Oxford University Press.

Hamilton, A. (2003). *The Federalist Papers*. New York: Signet Classic.

Illing, S. (February 6, 2020). "'Flood the Zone with Shit': How Misinformation Overwhelmed Our Democracy." *vox.com*. Retrieved from: www.vox.com/policy-and-politics/2020/1/16/20991816/impeachment-trial-trump-bannon-misinformation. Accessed January 31, 2020.

Klein, N. (2014). *This Changes Everything: Capitalism Versus the Climate*. Toronto: Knopf Canada.

Mann, G., and Wainright, J. (2018). *Climate Leviathan: A Political Theory of Our Planetary Future*. London: Verso.

Oreskes, N., and Conway, E.M. (2011). *Merchants of Doubt: How a Handful of Scientists Obscured the Truth on Issues from Tobacco Smoke to Climate Change*. New York: Bloomsbury.

Plato. (1955). *The Republic*, translated by Desmond Lee. London: Penguin.

Popper, K. (2013). *The Open Society and Its Enemies*. Princeton: Princeton University Press.

Searle, J.R. (2003). *Intentionality: An Essay in the Philosophy of Mind*. Cambridge: Cambridge University Press.

Thiele, L. (2019). *The Art and Craft of Political Theory*. London: Routledge.

Thucydides. (2013). *The War of the Peloponnesians and the Athenians*, edited by Jeremy Mynott. Cambridge: Cambridge University Press.

5
AUGUSTINE: LOVE

The first time I spoke in a church about climate change was very strange for me. Although the speaking invitation was from Anglican climate activists, the talk was advertised as interfaith and there were lots of indigenous folks there as well. It all went smoothly, but as the event progressed, I could not get out of my head the image of myself—someone who's bet that Pascal's coin is going to turn up tails—speaking from a pulpit to a group of theists, an enormous crucifix suspended from the wall just behind my head. I think the experience may have been technically dissociative: me floating above my own body and gazing down with a mixture of amusement and horror at the unfolding spectacle. But I could be making that part up.

Since then, I have given several talks about climate change to church groups. What I have learned is that when it comes to doing something about this issue these are the most indefatigable allies one could want. But there's more to it than that. They also understand the concept of the Anthropocene with a depth the religiously unschooled often can't grasp. What I mean by this is that they display a willingness to discuss the issue in the comprehensive way it demands. They understand that it challenges the way we think about human nature, the rest of nature and the relation between the two. They are ready-made big picture thinkers.

But what impresses me most about these folks is that not a single one of them has ever seen fit to ask me whether or not I actually believe in God. That itself speaks volumes about what they believe is and is not important in this struggle. I've therefore been led to ask what it is about religiously inspired climate change activists that makes them so refreshingly on-point about these challenges. Almost all my experience, both personally and professionally, has been with Christianity of one or another stripe. But the point I am making extends

across religions. Almost every religious community has released an official statement on the importance of fighting climate change: Baha'i, Buddhist, Christian,
Hindu, Interfaith, Jewish, Muslim, Sikh, Unitarian Universalist and more.

Let's take one of these, more or less at random: the Buddhist Declaration on
Climate Change, endorsed by 5,000 practitioners as well as 15 global Buddhist
leaders including the Dalai Lama. It reads, in part:

> As signatories to this statement of Buddhist principles, we acknowledge the
> urgent challenge of climate change. . . . We have a brief window of oppor-
> tunity to take action, to preserve humanity from imminent disaster and to
> assist the survival of the many diverse and beautiful forms of life on Earth.
> Future generations, and the other species that share the biosphere with us,
> have no voice to ask for our compassion, wisdom, and leadership. We must
> listen to their silence. We must be their voice, too, and act on their behalf.
> *(quoted in Aiken, 2015)*

Each statement says much the same kind of thing, and has the same broad-based
support from the relevant religious community. In other words, if you think
about all the world's religions together, a whopping 73% of humanity—about
5 billion souls!—are adherents of a worldview that calls explicitly for strong
action on climate change.

What is it that unites people of such diverse social and cultural circumstances on this issue? One phrase: care for creation or the biosphere. That's
certainly what I hear over and over again from theists. Whole books have been
devoted to exploring how each of these religions articulates this powerful idea.
I'm no expert in comparative religion, so I'm going to dig more deeply into
the one religious tradition I know the most about: Christianity. More specifically, I want to examine the thinking of one of the philosophical founders of
Christianity, St. Augustine. Why? Because he foregrounds the role that love
plays, or should play, in our understanding of how we fit into wholes beyond
the self. He does not talk at all about ecology, but this idea is in fact central to
what we now call ecological awareness. This is the sense in which Augustine
takes a big step beyond Plato. All of this will come out in the next few pages.
First, the crisis.

Barbarians within the gates

It is late August, 410 CE in what St. Jerome (347–429) called 'the mother of all
nations,' Rome. For three years, the barbarian forces, led by Alaric, a Christian
Goth (yes, these invaders were *both* Christianized *and* considered by Christian
Romans to be barbarians), have been camped in a ring around the city, with
troops stationed at all 12 of its gates, slowly choking the life out of it. They patrol
the Tiber river, the key source of goods into and out of the capital. Inside the

gates, people have become desperate. Here is Edward Gibbons' lurid account of life inside the walls during the siege:

> A dark suspicion was entertained that some desperate wretches fed on the bodies of their fellow-creatures whom they had secretly murdered; and even mothers . . . are said to have tasted the flesh of their slaughtered infants! Many thousands of the inhabitants of Rome expired in their houses, or in the streets, for want of sustenance; and as the public sepulchres without the walls were in the power of the enemy, the stench which arose from so many putrid and unburied carcasses infected the air; and the miseries of famine were succeeded and aggravated by the contagion of a pestilential disease.
>
> *(1985, 204)*

By the time the attack begins, the city's inhabitants are in a state of severe deprivation and trauma. The Goth armies roll over them, the culminating event in the fall of the Late Roman Empire.

At this time, the Empire was a mass of somewhat loosely stitched-together pieces. The most fundamental split, of course, was between the Eastern and Western empires, each containing multiple provinces that were, depending on their geographical and cultural distance from the two centers, more or less difficult to govern. This is hardly surprising for a political entity that at the height of its power stretched for millions of square kilometers and absorbed roughly 20% of the world's population. By the beginning of the 5th-century its influence and extent had waned relative to that pinnacle (achieved around 117 CE), but it was still a vast, internally seething colossus.

But at the pinnacle, life must have been truly grand, at least for the well-heeled. As Gibbons says about Rome in these glory days:

> If a man were called to fix the period in the history of the world, during which the condition of the human race was most happy and prosperous, he would, without hesitation, name that which elapsed from the death of Domitian to the accession of Commodus.
>
> *(1985, 134)*

That, to be precise, is between 96 CE and 180 CE, a period culminating in the death of the famous emperor and philosopher Marcus Aurelius and known as the Pax Romana. One of the period's most famous orators was Aelius Aristides (120–129 CE). Aristides' Roman Oration is similar to Pericles' Funeral Oration in illuminating the self-understanding and pride of a people at the height of its power.

For Aristides, the most remarkable feature of the Romans of this period was the harmony they brought to a disparate political hegemon:

> But the most notable and praiseworthy feature of all, a thing unparalleled, is your magnanimous conception of citizenship. All of your subjects (and

this implies the whole world) you have divided into two parts: the better endowed and more virile, wherever they may be, you have granted citizenship and even kinship; the rest you govern as obedient subjects. Neither the seas nor expanse of land bars citizenship; Asia and Europe are not differentiated. Careers are open to talent. . . . Rich and poor find contentment and profit in your system; there is no other way of life. Your polity is a single and all-embracing harmony.

(quoted in Oliver, 2006, chapter II)

There is no other way of life. A totalizing political and cultural system. Again, this is quite some time (230 years) before the Empire's fall. But the feeling of superiority and overwhelming might extended for a long period of time on either side of the historical window defined by the Pax Romana. Even though disintegration had set in after the death of Marcus Aurelius, there was still a feeling among many Romans themselves at this time that they occupied a high point in the history of the world, as was the case with the Athenians just before the war that destroyed their army and empire.

Focus for a moment on that notion of harmony. We can allow that Aristides is likely overstating the case for it but, even so, it tells us something important about the way the Romans saw themselves. What they perceived was a universal empire, one in which a staggering variety of disparate racial and ethnic elements had managed to come together under the political and military sway of Rome. There was, in other words, a feeling of unity and a sense of common purpose. Again, there's no reason to take any of this at face value, but at the very least it does constitute a cultural and political aspiration. For our purposes, what matters is not the reality of harmony within the Empire but the belief in it, however aspirational that belief might have been.

Notably, this feeling extended well into the Christian era. Beginning in 313 CE, the Empire gradually became Christianized. Constantine I legalized Christianity in that year, and in 379 CE Theodosius I made it the official religion of the Empire. Constantine pushed the boundaries of the Empire relentlessly. It is assumed by some historians that he converted to Christianity for purely prudential or political reasons, but whatever the reasons he did manage to consolidate this religion's hold over the sprawling entity.

Of course, we should not imply that the Empire at this time was a simple extension of what it was during the Pax Romana. The "magnanimous" ideal of citizenship celebrated by Aristides in the previous quote gave way in time to a more ruthless Christian imperialism. And Constantine himself was not exactly a choir boy, having murdered his wife, Fausta, and their oldest son, Crispus. Christian or not, emperors will be emperors I guess.

So the city Alaric and his troops entered in 410 CE was officially Christian, and had been for more than a generation. But this designation conceals as much as it reveals, for the peoples on both sides of the gates at that time. In spite of the feeling of common purpose and destiny, Roman culture was not uniformly

Christian in any sense. Pagan ideas were still very much in the air, especially among the nobles, and, as we will see in a moment, those voicing them tried to hold the Christians to account for the fate that befell the Empire.

In addition, the Romans had been assimilating barbarian peoples for a good 500 years prior to 410 CE. These groups were permitted to settle in Roman lands in exchange for military service. By the late 4th-century CE, notes historian Kyle Harper, "the barbarization of the army" was in full swing (2017, 193–194). One predictable result of this process, as we have noted, is that many Goths became Christians, and in this way Christianity penetrated into northern Europe, a part of the world that had staunchly and proudly resisted Roman incursion for ages.

According to Harper, an underappreciated factor here was that the surging Goths themselves began to feel pressured and threatened by Hun forces to the east. For centuries, the Huns had occupied the Eurasian steppe, a zone stretching from the Danube in the west to Mongolia in the east. They were the fiercest cavalry warriors ever seen, and were armed with the deadly composite reflex bow, a weapon with a range of nearly 150 meters. At the time in question, however, their eastern lands were ravaged by climate change-induced drought. The Huns, says Harper, "were armed climate refugees on horseback." They pushed ever further south and west, right into territory occupied by the Goths. In 376 CE, more than 100,000 Goths appeared inside Roman territory in desperate flight from the terrifying Hun invaders (Harper, 2017, 193–195).

The Goths surging into Rome are thus themselves being pressured from behind, a nice reminder that geopolitical events like this admit of multiple levels of causal analysis. In any case, the ultimate upshot of the Goth invasion is that the Empire shrinks dramatically. Far-flung provinces like Britain are basically left to fend for themselves. But this implosion does not come with the sense of a hopeful gathering together of forces that retrenchment can sometimes bring. Harper goes on to argue that the key feature of this new reality, for the Romans, is a sense of moral despair and fracture. That is certainly understandable. After all, "the native Roman population outnumbered the new arrivals, but the barbarians commandeered the superstructure of the state" (Harper, 2017, 195). At the very center of the realm the Romans had become a homeless people.

As we saw in the previous chapter, the Athenians too were a beaten people by the end of the Peloponnesian War. But their city was still, in some sense, theirs. The enemy was always now to be considered at the gate but with vigilance the city's moral and political topography might remain more or less recognizable. That they needed to work tirelessly to keep things that way, or at least prevent them from degenerating any further, partly explains the action taken by the Athenians against Socrates. But the Goths were not at the gate, they were *in the house*. Worse yet, they were running the place.

So a good deal of the estrangement and disorientation felt by Roman citizens at the time must have been about being forced to live cheek by jowl with those they had always considered barely human. Invasion is one thing, but *occupation* by those one takes to be ethnically inferior is an especially galling state to endure.

It is this living presence of the despised Other in their midst that causes so much distress to the Romans. Assuming Aristides is at least partly correct about the sources of Roman pride—specifically, the sense of unity and harmony among the distant and disparate parts of the Empire—why should the presence of the Goths among them have perturbed the Romans so deeply? After all, and to repeat, these were Christians. The putatively harmonious Empire was evidently still deeply fractured on racial and ethnic lines. How could the Romans have allowed this unwelcome incursion?

The bishop of Hippo

The Roman historian Sallust had already written the definitive account of the fall of the earlier Roman Republic (which lasted from 509 BCE to 27 BCE). The chief cause of *that* disaster, he argued, was moral. In particular, Rome had abandoned traditional values of civic virtue and allowed material greed and the unfettered pursuit of power and fame to take its place. More than four centuries on, this general account of political decline was still authoritative for many Roman citizens (Brown, 1969, 311). In the present circumstances it became all too easy for these pagan Romans to identify the source of moral corruption: the Christianization of the Empire at the hands of Constantine, Theodosius and the whole line of Christian leaders and apparatchiks following in their wake. As with the demise of the Republic, on their watch the decline of civic virtue had brought catastrophe and shame to a proud people.

Defenders of the Christian order thus had some explaining to do. As Augustine biographer Henry Chadwick puts it, "if St. Peter had replaced Romulus as patron of Rome, what was he doing in 410?" (2010, 127). It is no surprise that Romans turned beseechingly to the Empire's most brilliant Christian thinker, orator and polemicist to provide the explanation. Augustine became bishop of Hippo (now Annaba, a seaport on the northwestern coast of Algeria) in 375 CE. In his *Retractions*, he appeared eager to take up the challenge of defending Christianity against the pagan challenge after the fall:

> Rome having been stormed and sacked by the Goths under Alaric their king, the worshippers of false gods or pagans, as we call them, made an attempt to attribute this calamity to the Christian religion, and began to blaspheme the true God with even more than their wonted bitterness and acerbity. It was this which kindled my zeal for the house of God and prompted me to undertake the defence of the City of God against the charges and misrepresentations of its assailants.
>
> *(quoted in Ferrari, 1972, 206)*

We'll come back to this defense in the next section of this chapter. First, let's say a bit more about the defender. In his youth, Augustine was a Manichean, a form of paganism with deep roots in Persian and Indian religions. The Manicheans believed, among other things, that the universe is deeply split between opposed but equally balanced forces of good and evil. Evil, on this view, is a real thing, an entity whose activities in the world could be felt tangibly in all the terrible things that happen to us.

Augustine was a Manichean for nine years, before turning officially against these ideas. This is important to note because, in my view at least, he never really abandons the radical moral dualism that defines the thinking of this sect. We can trace Augustine's official rejection of it to two massive intellectual discoveries or epiphanies in his life. The first was his move, in 384 CE, to Milan to take up an appointment teaching rhetoric. In Milan he discovered the writings of the Neoplatonist philosophers Plotinus and Porphyry. Two years later, after coming under the influence of Ambrose, the bishop of Milan, he converted to Christianity. In defining Augustine's mature philosophical vision as well as his attempt to explain the fall of Rome, these two influences should be taken together. Christianity and Neoplatonism come to define his philosophical world.

Take the Christianity first, especially its way of understanding one of Augustine's perennial concerns, evil. Augustine's *Confessions*, the world's first real autobiography, is a remarkable document. Even for non-Christian readers encountering it for the first time, it can come across as an endearingly intimate self-portrait. But it's really Christian metaphysics *as* autobiography. Augustine characterizes himself in his youth as a sensualist and *bon vivant* as well as a determined social climber seeking all the material trappings of literary fame. He had always had a sense that the life he was living was not right, morally speaking, but Manichean philosophy gave him a way to rationalize it. After all, even if he was mired in a life of sin, the fault was not his own. He had fallen, unluckily, under the sway of the great cosmic force of evil.

Augustine writes the *Confessions* as a testament to his belief that this evasion is no longer open to him. Why not? Because on the Christian conception, there cannot *be* a quasi-divine force equal in power to God. To believe otherwise contradicts God's omnipotence, a key element of Its essence. Furthermore, if God is, as the Bible makes It out to be, the creator of all, how can It have made a part of the universe into something positively evil? Would that not make God evil? It seems as though God must *contain* evil to have created it. But God is the summation of Goodness. On the Manichean view, therefore, this positive evil force would constitute a *limit* on the real divinity, a limit either of power or goodness. To put the point succinctly, the Manicheans espoused a simple contradiction: they placed limits of goodness and potency on a Being that is, by definition, devoid of such limits.

So how do you explain evil? The explanation that emerges at this time and that is developed and refined for centuries thereafter is that the only real evil in the world is 'moral evil,' the product of free human choice. This sounds more straightforward than it is. To see how the claim works, we need to appreciate the other intellectual current behind Augustine's turn, his Neoplatonism. Plato had inaugurated a tradition of thinking that culminated in the doctrine of the Great

Chain of Being. Reality is, on this conception, composed of a rationally ordered hierarchy of types, ascending from minerals, plants and animals to humans, angels and God. Each level of this ladder is separated from the level above it by a metaphysical glass ceiling. In turn, each individual thing within each level has a specific set of functions, skills and capabilities.

This order allows us to be clear about what we should and should not expect from those individuals. The Chain is bifurcated directly across the middle. The top half is the realm of Being, the bottom half that of Non-Being. Full Being is God. Humans occupy a special place on the Chain because we straddle the two realms, Being and Non-Being. This, it turns out, makes us very difficult to control because we are drawn both downwards and upwards. Away from God and towards God. This is the key to grasping why we sin: we are pulled down, although we really want to go up.

The salient distinction behind this metaphysical picture is between (a) something we lack and should *not* have, and (b) something we lack but *should* have. The first is called a negation, the second a privation. To understand negation take clown fish and the ability to do calculus. We can imagine an especially disgruntled and angst-ridden clown fish moping around all day, wishing she could do calculus. 'This isn't fair,' she might grouse, 'I spend nearly all my time in a school, and I still can't do calculus. Something is seriously wrong with the system!'

I have a lot of sympathy for this fish, but clearly she is confused. Her inability to do calculus is, the Neoplatonist philosopher will say, a mere negation. That means that she lacks something that does not belong to the kind of thing she is. By feeling deprived because she can't do calculus our fish is supposing it is possible to break through the glass ceiling of her type. This poor fish desires a metaphysical impossibility.

But privation is different. When Augustine was just 17, he moved for a time to Carthage, in what is now Tunisia. Carthage sounds like it must have been a lot of fun at the time, at least compared to the sleepy and uptight suburb I was stuck in as a teen. Augustine describes it as a "hissing cauldron of lust," which is the very last thing you'd say about my hometown. In any case, he immerses himself into the cauldron with unbridled enthusiasm.

The philosophical story he tells about this descent is deeply moving. The power or capacity which draws him into the sensual is not in itself bad. Indeed, it is something intrinsically good: the will in search of something to love. In his case this will is basically misdirected:

> I began to look around for some object for my love, since I badly wanted to love something. I had no liking for the safe path without pitfalls, for although my real need was for you, my God, who are the food of the soul, I was not aware of this hunger. . . . So I muddied the stream of friendship with the filth of lewdness and clouded its clear waters with hell's black river of lust.
>
> *(1961, 55)*

Augustine's failure to turn his will towards its true object, God, is not a negation, like the clown fish's inability to do calculus. It is rather a privation, the absence of an orientation he ought to have. But though it is often damnably difficult to do, the will *can* turn in the appropriate direction. This is crucial: we are free to choose which paths to go down, and so the responsibility for our choices is ours entirely. If we do bad or evil things, that's not on God.

The whole text of the *Confessions* is meant to prove this point. Augustine thinks his *life* proves it. The nature of freedom, and of the will more generally, is a philosophical problem of dizzying complexity. The good news is we can set it aside and simply notice that in the process of laying out the details of his moral and philosophical journey Augustine provides a powerful model of moral trans-formation, one that is applicable well beyond the Christian, or even religious, context. According to this model, the quest for moral perfection is all about figuring out what and how to love. Before saying more about this, however, let's quickly note two problems with the distinction between negation and privation.

First, it remains difficult to understand the presence of defects in individuals that ought not to have them. Our clown fish can't rightly complain about the sorry results of her efforts to do derivative proofs, but she'd have a legitimate gripe if she could not swim fast enough to evade eels and sharks (though in this case, of course, her griping days would be short-lived). In other words, although the account on offer can, just maybe, explain moral evil (sin, etc.) it has a tougher time with 'natural evil.' And the category of natural evil includes anything from defective plants and animals to natural disasters like volcanic eruptions that kill lots of people.

Second, even the explanation of moral evil as privation is wanting. For we can always ask why a good God would have made us with such a wobbly will. Augustine's answer to this question is to invoke original sin. We have inherited this condition from those arch-sinners, Adam and Eve. This is the genetic con-ception of sin. Along with the genes allowing us to navigate our environments more or less smoothly, our parents have passed us this fateful capacity for evil. I have never heard a convincing account of how this is supposed to work, prob-ably because it is objectively silly. The more interesting issue has to do with the central place of love in Augustine's philosophical vision. Let's talk about that for a little while.

What's love got to do with it?

In one sense love is the simplest thing in the world. Everyone 'knows' it when they feel it. But this emotion is very difficult to understand fully. We can't hope to do justice to the wealth of thought devoted to exploring it over the ages, so we'll structure this section by posing three questions about love. First, how can we define it? Second, what is it about beloved things that makes them lov-able? Third, why do we love? The dialectical waters will get a bit choppy, but we will come out of the struggle with a better appreciation of why love matters in the climate crisis.

The first question is the easiest of the three. As with other concepts explored in this book I'll try and be as ecumenical as possible here, offering a definition of it that just about anyone ought to be able to endorse. Central to the idea of loving is surely the notion of adopting an attitude of caring towards the object of one's love. And to care for this object is to treat it with compassion, advancing its interests for their own sakes. That last bit is crucial. I cannot be said to truly love someone or something if I advance its interests only because doing so advances my *own* interests. That seems adequate for now. Love is an attitude of care a lover adopts towards an object, the beloved, so that the lover is inclined to advance the interests of the beloved for their own sakes.

That brings us to the second question, a far more complex one. What might prompt any would-be lover to take up this attitude towards some object? That is, what is it about the object that might cause someone to love it? Does it possess 'lovable' qualities? What might they be? Here it looks as though we can give two fundamentally distinct kinds of explanation, each one of which corresponds to a certain understanding of love that has been articulated throughout our history.

The first kind of love is based in the Greek concept of *eros*: an object is worthy of love to the extent that it contains or exhibits a certain valued property. When asked why he loves Susan, Charles says it is because of her angelic smile and wicked sense of humor. Queried about the basis of her love for her country, Angela tells us that it has a robustly democratic political culture as well as the world's best beaches. Presumably, absent these qualities—smile, humor, political culture, sand and surf—the loves in these two cases would not arise. And the same appears to be true for any other kind of love. We might think this is the best answer to the question we have asked about what makes the beloved lovable.

And so it was in the history of the West until about the late 4th- to early 5th-centuries. And then Augustine claims to find in the Bible, specifically in the writings of St. Paul, a qualitatively distinct kind of love based on the divinity's love for humanity. This is *agape*, also a Greek word. The idea here, according to some, is that God loves us precisely *because* we are lacking in any valuable properties or qualities. Here's the best way to think about this, a way that marks agape off from eros nicely. With eros, the attitude of care comes after the discovery of the relevant property or properties in the object. It is a response to this discovery. With agape, the attitude of care comes first and never seeks out a property or properties to corroborate it. For Augustine, this is the nature of God's love for us, and we should emulate it in our treatment of other humans. Do you doubt that this sort of love is applicable beyond the Christian context? If you are inclined to answer in the positive, an example might change your mind.

The philosopher Harry Frankfurt argues that agape is the model for *all* genuine love. For him, the paradigm form this takes is not God's love for humanity but rather parental love. Here is what Frankfurt says about this:

> The particular value that I attribute to my children is not inherent in them but depends upon my love for them. The reason they are so precious to me is simply that I love them so much.
>
> *(2017, 229)*

Another way to put the point is to remind ourselves of how often we say that the love we have for our children is *unconditional*. To say this surely implies that it is not the presence of some valuable quality in the child that serves as the basis of our love, but that the love precedes all conferral of value. Here we have a fully secularized version of what emerged in our history as a specifically Christian concept. All of this sounds difficult to deny. And yet, as the philosopher Alan Soble has pointed out (2017), we should be suspicious of the account. Inspired by Soble's critique of Frankfurt, let's take a more critical look at agape.

In calling parental love the paradigmatic form of this emotion, Frankfurt is saying that all loving should aim at the agapic paradigm, the feeling of love coming *before* the recognition of valuable properties in the beloved. All love is best thought of as aspiring to unconditionality, even if in practice it falls short of this ideal. But that's surely implausible. It does not seem applicable to romantic love, for example. Charles, recall, loves Susan just because of that smile and sense of humor. Perhaps he had always been drawn to women with just such features, whereas women lacking them left him cold. How do we explain this on Frankfurt's view?

Or take the case of patriotism. If Angela's country lurched to authoritarian rule and its beaches became inundated by rising seas, would she continue to love it? Maybe, but what if she judges that these alterations look to be permanent, so that the features of her country she once loved were, by her own reckoning, erased? My hunch is that she might cease to love her country in this case, or that at any rate she should. If she continued to feel love for it, the feeling might be based on nostalgia for what it once was, or perhaps hope against hope that things will change for the better. But if pressed, I suspect she would find it difficult to say that she loves her country as it is *now*.

These examples push us towards the view that all love is a form of eros, not agape. It is always about seeking out properties in the beloved, the recognition of which can then inspire the feeling. I don't want to oversimplify this. For one thing, it's surely not wise to dictate *which* properties are properly lovable. When it comes to who or what we should find lovable I'm a pluralist. Let a thousand erotic flowers bloom. Even so, we need to do more work establishing the claim that all love is erotic. After we do so, we'll revisit Augustine.

Let's return to the hard case, parental love. Is it really unconditional? Frankfurt says that he loved his child even before it was born, before he could be aware of any properties it might have. But that's because he assumed it was going to be a roughly recognizable instance of its kind, a little bundle of *Homo sapiens* sweetness. What if he and his wife conceived the baby in space and that, unbeknownst to either of them, the mother had contracted a virus resulting in the implantation in her of one of those horrible monsters from the *Alien* movies? There's a sense in which the thing is still their child, but surely it's unlovable. In any case, since it's going to eat them it would be prudent of them to kill it quick. In that terrific 1976 film *The Omen*, the child Damien—the devil's progeny, as it happens— takes delight in provoking violence and general mayhem. He is also incapable of displaying the smallest degree of kindness or sympathy to his parents, and that causes them to *cease loving him*. Can being radically unloving make one unlovable? I don't see why not.

I saw that film when I was 11, and I remember thinking that maybe my parents' love was, in fact, conditional on my good behavior. Better be a bit nicer, even if my nastiness wasn't quite on a par with that of the spawn of Satan. So what's really going on with Damien? At one point in the story, a peripheral character says, "That child is not human." There's our clue: Damien lacks a valuable quality or property—just the minimal requirement of being human, but also certain affections that generally come with that—the absence of which robs him of value and thereby renders him unlovable. Frankfurt claims that he would still love his children even if they became, like Damien, "ferociously wicked." Not me, and yet I suppose I love my children as much as he does his. Even here, in that relation that is supposed to be the quintessence of love-as-agape, we therefore find eros.

So we have no good reason to follow Frankfurt, or Augustine, down this particular path. However, Augustine is absolutely right to have placed love at the center of his understanding of the moral life, and that is why his account is still so valuable. The emphasis on love allows us to understand how moral progress actually works: by giving us reasons to act caringly towards its objects. We come to see that a certain group, or class of entities, has interests that we should care morally about. In theory, the point applies as much to slaves as to endangered species.

In other words, moral progress occurs through our challenging dominant claims about *which* properties count as valuable and lovable. Centering love reveals what drives the process of ever-expanding inclusivity. Advances in moral progress always consist in demonstrating that a certain group—the oppressed, the outsiders, the despised, the different—possess properties that make them morally considerable. Deciding to care for someone or something marked out this way then engages the lover's agency in a specific way: to advance the beloved's interests.

This brings us, finally, to the third organizing question of this section. Why do we love? We have claimed that love involves care for the beloved and that it is essentially erotic—the answers to our first two organizing questions—but why should we not shun its messy demands altogether? What are we seeking in indulging it? As we have seen, for Augustine love involves a turning of the will towards its proper object, God. In the Neoplatonic philosophical outlook this is a turn towards our original ontological source or home. In this tradition, moral enlightenment is described as a loving ascent up the Great Chain of Being.

The philosopher Simon May provides a secularized version of this idea. He thinks the quest for home is central to all love's forms. For May, love is a response to a feeling of existential exile or disorientation. We fasten on this or that object of love because doing so holds out a promise of security and sense of homecoming. Love, he says, "names our joyful response to a promise we glimpse in another to meet this need for rootedness or groundedness or home" (2019, 41). As such, love has a morally *revolutionary* capacity:

> Since love thrusts into view . . . a new sense of home, it uproots us from where we are, from how we live, in order to root us more fully in the place towards which the promise beckons. In order to (re-)discover our true home we must lose our accustomed habitat—our habitual mode of living. We must, in a certain sense, experience ourselves as being displaced and even in exile.
>
> *(2019, 42)*

The insight is basically Augustinian and has no precedent in the history of philosophy prior to this thinker. More importantly, it supplies exactly what was missing in Plato: a reason to care about the larger wholes in which we find ourselves, *motivation* for protecting what is valuable.

For if it is true that the Anthropocene is best characterized as an age of displacement and exile, we can now understand the full depth of this event. It can sound so anodyne to say that we are transitioning from one geological epoch to another, especially if we state this claim in the austere terms of stratigraphical science. But if we can also begin to articulate the complex thought that we have lost a home, that the loss is due to our own history of misdirected willing, that this has caused a deep sense of exile in us and that love might illuminate the path to a new home, we will have said so much more about the existential labyrinth in which we are trapped.

We could not have couched the matter this way, however, absent Augustine's novel understanding of love's role in our communal lives. Moreover, although he articulates the view in his writings before the fall of the Empire, that event adds considerable historical and political depth to it. So now, at last, we are positioned to return to Augustine's answer to his compatriots.

Entanglement

He was, it seems, in no rush to produce the defense of Christianity they were seeking. The *Confessions* was completed by about 401 CE, but the *City of God* (Augustine, 1998), which contains the sustained defense, did not appear until around 427 CE. That's 17 years after Alaric and his forces had overrun Rome. The *City of God* builds on the account of sin as misdirected willing laid down in the *Confessions*. But in the later work, Augustine is much more pessimistic about our ability to escape the clutches of sensuality than he is in the earlier work.

This apology for Christianity must have been a bit disappointing for many readers. If they had been hoping for a story about how this religion could provide its adherents with a reason to believe that Rome, the eternal city, could rise from the ashes, no such account was forthcoming. Augustine does pour plenty of scorn on the pagan gods for *their* abject failure to protect the city. And he notes with some approval that the Goths, presumably owing to the constraints of conscience imposed upon them by their Christianity, did not engage in outright destruction of the city's temples or the wholesale slaughter of its citizens.

However, the book's overriding message is that Rome is not the eternal city after all but only the 'City of Man' (or 'Earthly City'), ruled inescapably by Satan. In the fullness of time it will give way to the 'City of God,' and so we should not wring our hands over its plight. In fact, should Romans not rejoice once they understand that these events are the necessary precursor to the time of fulfillment to come? And what is the City of God? It is, in the mind of this Christian, a utopia governed only by the laws of love. Love of God draws us towards it and love of each other reigns undisturbed within it. By contrast, down here in the realm of the senses, everything is awash in death, decay, finitude, ugliness and endless, pointless struggle.

There is a clear affinity between Plato and Augustine, an intellectual communion that is passed down to the entire Christian West and persists in one form or another right up to the present. They are both deeply otherworldly thinkers, and to the extent that we are still in the grip of their ideas there's an undeniable element of otherworldliness in our own metaphysical orientations (Lovejoy, 1936). Moreover, both Plato and Augustine become metaphysical dualists, or double-down on their dualism, in response to profound crisis. It is easy, and customary, to characterize the separation of the real into distinct spheres of the material and immaterial, Non-Being and Being, as a flight from reality. But what I've been urging here is the idea that it might also be just the opposite: a desperate attempt to find our place in a reeling world, a world of chaos, collapse, fear, disunity and the unavoidable face of the feared or despised Other.

In historical circumstances like the ones faced by these thinkers, it can make sense to think dualistically, to come to believe that our true home is somewhere radically beyond this confusing place. But where exactly is it? Here is where Plato and Augustine part company sharply. In the *Confessions*, Augustine describes his dying mother, Monica, as "falling asleep" in God. Death is the literal *return* to a loving parent. It is a spiritually cozy habitation. Plato could never have believed something like this. The Forms do not exist in a *place*, let alone a tranquilly ethereal city. Just as important, Augustine's city is a place that in principle *anyone* can enter.

To be sure, God is an exacting and arbitrary gate-keeper, and original sin is a powerful downward force. So only an elect few *do* manage to get in, but membership is not restricted to an intellectual elite. Access to the realm of the Forms is, by contrast, tightly restricted to a certain type of person. We see here the truth in Nietzsche's description of Christianity as a "Platonism for the people."

Is there anything to this story for anyone not already convinced by the truth of Christianity? Indeed there is, and we can see this clearly by focusing on the term Augustine opposes to love, *cupiditas*. Cupiditas is desire moralized. There's no reason to besmirch desire as such. None of us would ever *do* anything—feed ourselves, procreate, cultivate our musical talents, exercise, rinse our vegetables, flee that onrushing cougar, etc.—in the absence of desire. But desire has a tendency to degenerate into greed, seeking out that which we do not need.

The real problem goes beyond taking what is not strictly required, however. It has rather to do with taking what others need *more* than you do when you can't both have everything you want. In the philosophical tradition, greed is generally opposed to justice, and it's with the contrast between these two attitudes—the first a vice, the second a virtue—that things get more interesting. To be just is to be inclined to distribute benefits and burdens equitably in a situation of scarce resources. The greedy person will do this inequitably. In other words, the just person has an eye to the welfare of the social whole of which she is a part, while the greedy person, also a part of that whole, wants only to advance her own interests.

Our plight in the Anthropocene cannot be fully comprehended without embracing this insight. We can, if we like, strip away talk of the City of Man as the domain of Satan. Although I like movies about the chap and his charming offspring, I have a difficult time imagining such netherworldly figures wending their mischievous ways through actual space-time. Augustine never really escapes his Manichean sensibilities, in spite of the evil-as-privation theory outlined earlier in the chapter. He could never quite shake free of the tendency to think of evil as some kind of mysterious *presence* in the world. I get that, because I too was taught to fear a personified devil. But I also got over it whereas I don't think Augustine ever did.

The core of the moral doctrine is nevertheless solid, because what Augustine emphasizes is the fact that down here the two cities are inextricably *entangled* and that the entanglement can be viewed as a contest among objects of love. Chadwick thus asserts, correctly, that for Augustine the "quality of a society can be seen by asking what it loves" (2010, 144). In real human history there is an ongoing battle between greed and justice, love narrowly directed and love broadly directed, care of the self and care of the whole. The human condition is one of irreducible ethical entanglement. I find this simple insight immensely clarifying, but what does it mean specifically for the climate crisis?

It means that we should be willing to name the forces in our culture, and in our own hearts and minds, that obscure the basic truth that care for the whole is the raison d'etre of structured collective life, i.e., of politics. The French cultural theorist Bruno Latour has said that "there is indeed a war for the definition and control of the Earth: a war that pits—to be a little dramatic—Humans living in the Holocene against Earthbounds living in the Anthropocene" (2015, 151). That *may* be a bit dramatic, but these are extraordinary times. And how else are we supposed to talk about those who use what the journalist Jane Mayer calls "dark money" to set up foundations aimed at deceiving people about the scope of the climate crisis?

Here for instance is Mayer discussing the sociologist Robert Brulle's research into what is happening with all this money:

> During the seven-year period [Brulle] studied, these foundations dis-
> tributed $558 million in the form of 5,299 grants to ninety-one differ-
> ent non-profit organizations. The money went to think tanks, advocacy
> groups, trade associations, other foundations, and academic and legal pro-
> grams. Cumulatively, this private network waged a permanent campaign
> to undermine America's faith in climate science and to defeat any effort to
> regulate carbon emissions.
>
> *(2017, 253–254)*

Although they get most of the press for these nefarious activities, the Koch broth-ers are just the tip of the gilded iceberg of American families, which also includes the Scaifes, the Olins, the Bradleys and more. Mayer's exhaustive research makes clear that the entire plutocratic class is concerned almost exclusively with advanc-ing its own interests, and that those interests are in most cases bound up with the perpetuation of the fossil fuel energy regime (among other things).

Here, then, the gritty journalist, the radical French social theorist and the old Christian saint come together. We have lost the battle to stop the climate change juggernaut wholesale because our politics has been intentionally dis-torted to favor the interests of the few at the expense of the interests of the whole. If this is not war, what would you call it? Resist the inclination to answer in the dry language of interests, as I have just been doing. How do you *feel* about bushfires of unprecedented ferocity that killed about 1/3 of the koala population in northern New South Wales in 2019–2020, even as Aus-tralia continued to export huge quantities of coal? I feel as though something eminently worthy of love, but also utterly helpless, had been destroyed for no good reason.

I'm talking here not about unconditional love but love-as-eros. What would it even mean to say we might or should love these beings unconditionally? No, the biosphere is lovable *because of* its qualities, its anciently beautiful and self-sustaining complexity. When I think of things this way, I also feel deep anger because our reckless evisceration of this web of wondrous things strikes me as a moral crime of world-historical proportions.

Augustine is particularly helpful in the current impasse because he allows us to see that, in cases like this, self-love and greed have triumphed over a more expansive love. Religious folks refer to the object of this expanded affection as Creation while atheists call it the biosphere or the Earth system (and the beings these systems contain). But even though it has ultimately to do with the very origins of the cosmos, from the standpoint of the climate crisis this difference between the two perspectives is objectively trivial. Both theists and atheists can, and many do, feel a deep and abiding *biophilia*, the love of life in all its evolved diversity.

What is biophilia? The term was coined by the biologist Edward O. Wilson (1986) to designate that feeling of affection so many of us have for nature. It encompasses a whole range of psychological states and attitudes towards non-human life, from the feelings of peace and contentment we derive from spending time in nature to our aesthetic preference for landscapes that are similar to our ancestral home, the African savannah. May points out that "we see in our loved one a powerful and very specific lineage or heritage . . . and so a source of life, with whose sensibility we deeply identify" (2019, 45). This can be applied to biophilia because what we are responding to with this emotion, so goes the hypothesis, are the bonds established over the course of a long co-evolutionary history with other living things. It also explains the deep sense of guilt and shame many are now expressing at our destruction of the biosphere. Talk of 'mourning' for what we are losing is now rife among those inclined to psychologize the issue.

Thinking about the Sixth Mass Extinction in terms of biophilia allows both theists and atheists to, for example, recognize Jair Bolsonaro's razing of the Brazilian Amazon, begun in 2019 to make space for ranch land, for exactly what it is: a brazen attack on the primal nurturer. Both can say sincerely that our times have pitted a greedy minority against the planet. There are no significant differences between the religious and non-religious on this score. Augustine is crucial both in helping us see the struggle as having mainly to do with misplaced love and in showing us that when it comes to speaking up and acting on behalf of a threatened loved one we should never be mealy-mouthed.

There's a twist here, one Augustine is also instrumental in exposing. Augustine does not exactly divide the world up into sheep and goats, as I've just been suggesting. The entanglement of the two cities also works its way into each of us. It's disarmingly easy to internalize the prerogatives of Holocene power structures even if you do not sit right at the top of them. This is because so many people, especially in the developed world, have benefited from these structures. The two loves—between self-interest and the interests of the whole—can also rend the self internally. And so, the battle for a humane Anthropocene needs to be as much about overcoming our own casual overconsumption as it is about overthrowing the plutocrats. Politics must be conducted at both levels simultaneously.

Conclusion

As I remarked in the previous chapter's conclusion, for all his unsurpassed brilliance Plato's view of humanity is too coldly rational. He can help us see why our politics should be constrained by knowledge, but it is always possible to ask why we should care about this. The answer surely cannot be, as it seems to be for Plato, that the alternative would be worse *for the knowers*. Augustine corrects this oversight in Plato's thinking by arguing, throughout virtually his entire career, that how we love is a central feature of what we are. Love is about the will, not the intellect (or reason). It concerns our striving to make the world a better place for the beloved, to make it a place where the interests of the beloved can flourish. Lovers create flourishing and protective spaces for what they love. Lovers are the true guardians of the whole.

There's one more point to add here about the connection between care and creation. Humanity has always flirted with the idea of imitating God or the gods in their creative capacity. We've always been jealous of that presumed source of power. We see this recurring in some discussions of the Anthropocene. The environmental writer Mark Lynas recently wrote a book called *The God Species*, and, as we have seen (in Chapter 1), the title of one of Harari's books is *Homo Deus*. Everyone who talks this way now does so mainly in the sense that we humans have become uber-makers, just like that other Maker. The point, for them, is mostly about technology. And then there's the inevitable push back: 'playing God' is overly hubristic and will lead down the path of perdition.

Taking Christian thinkers like Augustine seriously can remind us that imitating God—if we insist on talking this way—should primarily be about *care-constrained* creation. This part of the equation is mostly missing among those who celebrate our new divinity. Combining this insight with the previous chapter's, we've now put in place a dual constraint on political decision making in the new epoch. The first is a constraint of knowledge, derived from Plato; the second, a constraint of love or care, which we get from Augustine. Both, however, are directed at promoting the interests of the whole and thus overcoming the greed and myopia in which our politics is now disastrously mired.

But remember, Augustine might not have come up with any of this absent Rome's crisis of invasion and occupation. Truths as important as this are always hard won. Crises can come in subtler, but no less momentous, forms than those considered in this and the previous chapters. In the next chapter we jump ahead about 1,200 years, to the modern scientific revolution, to see how intellectual crisis can also prompt the overthrow of a whole metaphysical system.

References

Aiken, W., et al. (May 14, 2015). *The Time to Act Is Now: A Buddhist Declaration on Climate Change*. Retrieved from: http://fore.yale.edu/files/Buddhist_Climate_Change_State ment_5-14-15.pdf. Accessed April 25, 2019.

Augustine, St. (1961). *Confessions*. London: Penguin.

———. (1998). *City of God*. Cambridge: Cambridge University Press.

Brown, P. (1969). *Augustine of Hippo: A Biography*. Berkeley: University of California Press.

Chadwick, H. (2010). *Augustine of Hippo: A Life*. Oxford: Oxford University Press.

Ferrari, L.C. (February–March, 1972). "Background to Augustine's City of God." *The Classical Journal* 67, 198–208.

Frankfurt, H. (2017). "On Love and Its Reasons." In *Desire, Love and Identity*, Gary Foster (ed.). Don Mills: Oxford University Press, 228–233.

Gibbons, E. (1985). *Decline and Fall of the Roman Empire, Abridged*. London: Bison Books.

Harper, K. (2017). *The Fate of Rome: Climate, Disease and the End of an Empire*. Princeton: Princeton University Press.

Latour, B. (2015). "Telling Friends from Foes in the Time of the Anthropocene." In *The Anthropocene and the Global Environmental Crisis: Rethinking Modernity in a New Epoch*, C. Hamilton, C. Bonneuil and F. Gemmene (eds.). London: Routledge, 145–155.

Lovejoy, A.O. (1936). *The Great Chain of Being: A Study of the History of an Idea.* Cambridge, MA: Harvard University Press.

May, S. (2019). *Love: A New Understanding of an Ancient Emotion.* Oxford: Oxford University Press.

Mayer, J. (2017). *Dark Money: The Hidden History of the Billionaires Behind the Rise of the Radical Right.* New York: Anchor Books.

Oliver, J.H. (2006). *The Ruling Power: A Study of the Roman Empire in the Second Century After Christ Through the Roman Oration of Aelius Aristides.* New York: Kessinger Publishing.

Soble, A. (2017). "Two Views of Love." In *Desire, Love and Identity*, Gary Foster (ed.). Don Mills: Oxford University Press, 221–226.

Wilson, E.O. (1986). *Biophilia: The Human Bond with Other Species.* Cambridge, MA: Harvard University Press.

6

DESCARTES: THE TECHNOSPHERE

I love whiskey in pretty much any form it deigns to appear before me—scotch, bourbon, Tennessee, rye, Irish, you name it. At bottom, makers of great whiskey are skilled and fastidious purifiers. The process of distillation required to make hard spirits like these is a refined art of incremental separation. Whiskeys are made from grains—barley, corn, rye—mixed with water to create a mash. The process of fermentation gets to work on the mash, and the distiller then separates the resulting alcohol from the rest of the mixture through condensation in special vessels (pots or giant columns). But elements known as 'congeners' evaporate at the same time, tannins and esters that provide flavor and aroma to the spirit but can also make it unpotable if there are too many of them. Because the concentration of congeners in the final mix is so important, numerous iterations of the distillation process are required, sometimes 30 or more.

When I first observed it on a tour of a distillery just outside of Nashville, the distillation process struck me as a pretty good metaphor for a lot of things we do. Now, I'd had more than a single dram of fine Tennessee bourbon when the bulb lit up, but in spite of this, or perhaps because of it, I recall thinking that whatever else you want to say about us we humans are relentless purifiers. The general technique is ubiquitous, showing up in practices that range from the noble through the benign to the positively evil. Nothing wrong with great whiskey, of course, but when a white supremacist talks about outsiders sullying the purity of the race and advocates expelling them from the land, the application of the term—not to mention the associated political 'techniques'—has become sinister indeed.

That more sinister sense of the project of purification will come up again in the next chapter. In this one, we are going to look closely at the way modern science has purified the material world to make it pliable to our technologically defined purposes. We'll do so through an examination of the thinking of René

Descartes (1596–1650), an early-modern philosophical luminary. For Descartes, we have an irrepressible urge to understand our world, as well as transform it technologically, and yet at the time he is writing we are mired in ignorance about it. His job, as he sees it, is to rid this world of all the things—sensations and purposes—that stand in the way of mathematizing the whole. And he wants to do this so that we might become, as he puts it, its masters and possessors, which he thinks of as humanity's highest moral task. This is a fundamental turning point in our history. You might not like everything this vision has wrought, but it is difficult to disown sincerely.

To this day, Descartes' discoveries in this area are underappreciated. The reason for this is twofold. First, when we think of the revolutions of modern science, cosmology gets the lion's share of our attention. In declaring that the Earth moves around the sun and not the other way around, Copernicus and Galileo unseated centuries of dogmatism rooted in the false cosmology of the ancient Alexandrian astronomer Ptolemy (100–168 CE), a worldview later propped up by the Bible. I have no desire to downplay the significance of this discovery, but it's a pity that it has consigned to the shadows a no less revolutionary alteration in our understanding of the natural world.

Second, Descartes' claims about the fundamental structure of the physical world run entirely contrary to what common sense thinks about it. The Cartesian view is that matter is nothing but bits of atoms in motion, atoms known at the time as 'corpuscles.' Literally *everything* about how corpuscles behave can be understood mathematically. Bodies are nothing but temporary agglomerations of corpuscles whose relative size and motion can be measured. Sounds pretty innocent so far, right? I agree, but what's revolutionary is not so much what Descartes says is *in* bodies as what he says they *lack*.

For the dominant theory at the time held that sensory qualities—taste, tactile feel, smell, sound, to say nothing of values and purposes—are in bodies. Descartes' entire career was, in my view, devoted to expelling these things. In this chapter I'm going to argue that Descartes' whole philosophical enterprise is dominated by this quest to purify matter, a purification which has resulted in the creation of what has become known as the technosphere, a tangled congeries of artefacts covering the whole planet, currently weighing in at 30 trillion tons (Zalasiewicz et al., 2016).

Staggering though it is, even this figure underrepresents our impact on the planet because it excludes weightless technologies and their effects, like genetically modified organisms or the altered chemical makeup of the atmosphere. What crisis helped create the technosphere?

Paradigms, revolutions and idolatry

Crises do not have to take the form of wars, revolutions and invasions to shake a people thoroughly. There are intellectual crises too, crises of knowledge, and they can have a huge impact on the course of human history. My topic in this section is the nature of scientific paradigms: how they work, the way they define a way of being in the world, and the forces that can prevent their graceful retirement from the historical stage. In my view, given its role in literally *making space* for the Anthropocene, there is no more fundamental alteration of such worldviews than the modern scientific revolution.

It began towards the end of the 16th-century, gathering steady momentum throughout the 17th-century. It is no exaggeration to call it revolutionary even though this term gets bandied about far too casually. In physics, astronomy, chemistry and biology, our understanding of the workings of the natural world increased dramatically. And the insights extended well beyond natural science, encompassing social and political theory as well as ethics. How should we understand intellectual transformation on this comprehensive a scale?

Thomas Kuhn (1922–1996) changed our thinking about how science works with the concept of a "paradigm," a relatively closed field of directed scientific inquiry. Kuhn (1970) calls it a "disciplinary matrix," a social construct comprised of specific puzzles to be solved, relatively well-defined methodologies for solving them, a network of researchers, technologies and infrastructure, and an educational establishment producing papers, textbooks and new researchers. It is, in short, an engine for knowledge (re-)production and dissemination. Its piecemeal progress at puzzle-solving is what Kuhn refers to as "normal science."

Normal science can go on for a very long time while new discoveries are made and new professionals enter the knowledge-structure while others are circulated out of it. This is how any scientific discipline works for most of its history. Occasionally, however, more rapid breakthroughs occur, "revolutionary" leaps that set the discipline on a new trajectory of research and discovery. These are often resisted by defenders of the dominant paradigm. What immunizes normal science—at least partially—against fundamental change? And what, in turn, sometimes upsets this steady-state, allowing for the emergence of revolutionary thought-structures?

Kuhn cites the following experiment to illustrate the problem. Test subjects are shown a series of playing cards in quick succession. The subjects are then asked to remember the cards. They are all normal playing cards except for one. This is referred to as the "anomaly" card, say a black five of diamonds. For a large number of test subjects, the anomalous card is assimilated back into the old way of seeing things. For them, the black five of diamonds becomes, for example, a red five of diamonds. Gradually, as the subjects are exposed to more and more test runs, with more and more anomalous cards in the series, their perception of reality begins to shift. Some of them maintain stubborn adherence to the 'normal' way of seeing the cards, others eventually submit to the new perceptual reality.

Interestingly, many of those making the shift experience emotional distress at doing so. Evidently, the old card-seeing paradigm provides some degree of familiarity and comfort to people that goes beyond a merely intellectual pattern-sorting exercise. Subjects are emotionally *invested* in the normal way of perceiving reality (Kuhn, 1970, 62–64). The switch is a mini-crisis for the test subject.

Science works the same way, and this is why scientific revolutions take the shape of genuine collective crises. A scientific paradigm is a worldview, a comprehensive way of structuring our relation to the external world. I believe we should take very seriously the sense in which we can invest so much of ourselves in this way of seeing, and that's because the way of seeing is really a way of *being* in the world. The card experiment suggests that sometimes this way of being is confronted with contradictions it can no longer assimilate to the old way of data-processing. These are the anomalies.

No paradigm will ever be entirely free of anomalies. Their presence might indicate only that the science in question is incomplete, that there are some parts of the relevant data field it cannot yet explain fully. Normal science proceeds by way of trying to eliminate as many of these as possible by tweaking the theoretical structure so that it can account for them. But sometimes the paradigm will reach a point where there are so many anomalies or the existing ones are so persistent that a whole new way of thinking is required. These crises are the revolutionary moments punctuating the history of science, allowing it to set out on a new path. There's no formula for the transformation, for how jarring the anomalies must become before the whole thing collapses.

But while the card experiment *reveals* how readily we resist new information, it does not *explain* such resistance. We therefore need to look more carefully at the social, political and psychological forces that can prevent change. As it happens, the sharpest account we have of this is from the period under investigation. Sir Francis Bacon (1561–1626) was a statesman, diplomat, philosopher, author, jurist and legal advisor to Queen Elizabeth. As if those tasks were not enough to keep a person busy, some have even floated the idea that he was the real author of the works commonly attributed to Shakespeare. Whatever we make of that idea, which has been in circulation for well over 150 years (talk about a notion resistant to collapse!), Bacon's understanding of the forces behind the resistance to the new science is insightful.

In the *Novum Organum*, Bacon describes adherence to the old ideas as "idolatrous." That is, he thinks humanity is under the sway of certain idols which are as such getting in the way of understanding the workings of nature. Invoking an image like this is immensely powerful in a deeply religious age. Think of this at two levels. The first has to do with how our attention is focused. Idols, of course, are false gods, the worship of which supposedly turns us away from apprehension of the true gods or God. So in asking us to reject the idols, Bacon is instructing us about which way to turn the mind's eye.

Second, any object of religious devotion structures our whole way of being in the world. As with Kuhn's card recognizers, but presumably in a much more comprehensive way, it affects not just how we perceive the world, but how we relate to other people, our politics, how we understand our emotional makeup and so much more. So what were the idols impeding true science? Bacon lists four.

The first are "idols of the tribe," those allegedly inherent in the species. Bacon seems to have in mind here chiefly the tendency to draw conclusions over-hastily on the basis of scant evidence. Second are the "idols of the cave," which refer to our individual idiosyncrasies. These are the product of the unique education and upbringing each of us receives. Third, there are the "idols of the marketplace," the prejudices of our group encoded in the way we distinguish ourselves from others on the basis of our linguistic peculiarities. This describes those speaking different natural languages (like Spanish and Russian) as well as speakers of a single natural language which has fragmented into various dialects (like Parisian and Québécois French), but it also encompasses argots (like those of gamers, prisoners, teens, government bureaucrats, news pundits, etc.). Finally, we have "idols of the theatre," the political, religious or philosophical systems that bias the ways we see reality (Bacon, 1905, articles 41–44).

It's a rich conceptual structure. Taken together, it's easy to see how the idols can form a worldview. The more they work together, offering a coherent picture of reality to us, the more complete is the worldview. To put it in terms discussed in previous chapters, this coherence can offer a deep and comforting sense of homeliness to us. But it can be a problem when a new worldview, one perhaps closer to the truth of nature, is struggling to be born.

Bacon thought that all four idols were conspiring in this fashion to impede the emergence of the new science, arguing that the mind must be utterly liberated from their pernicious influence. Bacon is, of course, also the founder of what we now understand as *the* scientific method. That method would have us build our knowledge of reality from the ground up according to the method of "true induction." This is opposed to "syllogistic" reasoning or deduction. The basic difference between the two methods concerns the derivation of general principles or laws. Science cannot do without these, but the then–dominant paradigm begins with first principles that are, allegedly, self-evident. From there we can derive further principles or laws. Only after this exercise do we come to the world of particular things (planets and chairs, apples and emotions, etc.). To qualify as real, these things must fit what the theory says about them.

True induction, by contrast, proceeds by way of sensory apprehension of the particulars, constructing generalizations from there. The choice is stark: we either start with principles and bend the facts to fit them or start with the facts and construct theories on their basis (Bacon, 1905, articles 14–20). The distinction seems almost too cute and yet Bacon's advice was to have momentous consequences in the history of science. To take just one famous example, consider Newton's theory of universal gravitation. According to the dominant paradigm at the time, we need separate theories to explain what goes on with physical bodies above and below the moon.

According to this view, the basic push-pull principles of mechanics can explain bodily events in the "sublunary" sphere, everything below the moon. This is the domain of change and becoming, where earth, air, wind and fire are the key elements. Here, the old scientists accepted that there is *some* role for the senses to play in informing us about what is going on. But what went on above the moon, in the "superlunary" sphere of the heavenly bodies, was altogether different. In that celestial realm, relations among objects are eternal and can only be grasped by the first principles of geometry. The senses here are at best irrelevant and at worst completely deceptive. It's the sole job of reason, working with self-evident axioms, to enlighten us about how this place works. This is where God's mind roams.

Newton unified the two spheres, arguing that his theory of gravitation applied as much to falling apples as to the elliptical movements of the planets. It is a function solely of the quantity of mass each body contains. Just as important as this conclusion, however, is how he arrived at it. Nobody had yet visited the moon, after all, so how could we say for sure what went on up there, to say nothing of those bodies beyond the moon? Well, answered Newton, let's just stick with what we *do* know. We now have the mathematics to understand how the force of gravity works on bodies close to the surface of the Earth—that famous falling apple—so we should *infer* that it works everywhere else there are bodies too. This is the careful derivation of a generalization from observed particulars.

As it turned out, applying the principles of gravitation to the heavenly bodies allowed for much more accurate representations, and predictions, of their movements than the old science was capable of. Had he been around to appreciate it, Bacon would have admired and felt vindicated by Newton's innovation of fundamental physics. Careful inference was, for these philosophers, opposed to the construction of "hypotheses" independently of experience. And so Bacon would also have applauded Newton's way of defending his insight about gravity. In a barb aimed at the heart of the old paradigm, Newton asserts, "I do not feign hypotheses."

Together, Kuhn, Bacon and Newton give us a way of applying the concept of scientific revolution to the time that interests us here, the early modern period. We've seen what the revolution means for how we understand the basic laws governing the motion of bodies. That leaves open the question about what's *in* those bodies, the dominant question of Descartes' philosophy. To set it up, imagine the following casual conversation over wine.

The brain as factory

A chatty and philosophical friend shows up at your door bearing libations. You're not too busy this afternoon, the wine looks good, so you invite her in. After pouring the drinks, she asks you out of the blue how many things you think there are in the universe. You might want to dismiss the question out of hand, because it appears to be either hopelessly vague or impossible to answer. But you decide to play along. Unfortunately, your spirits take a plunge as you realize that the number of things in the universe is too big to count. Even a cursory look around the room the two of you are sitting in might convince you of this.

Besides the books, the armchairs, the dog, the stereo, the bowl of apples, the carpet, the paintings, and the wine glasses (among many other medium-sized objects), there are all the tiny things. Are we supposed to count every one of those dust motes dancing in the beam of light coming through the window? What about the beam of light? Seems like a thing, even though it pops in and out of existence depending on the play of clouds in the sky. Speaking of your dog, an exceptionally shaggy Alsatian, does every one of his hairs count as one thing?

What about the carpet fibers? While we're picking things apart this way, isn't each triangular pattern cut into the base of the crystal wine glasses something that is, in a way, distinct from the wine glasses themselves? And, finally, we shouldn't forget the pain in your knee from your morning run, the itch in your left earlobe or the inexplicable melancholy that washes over you about this time every afternoon. Sure, these things are 'just in your head,' but your head's in the room, right?

This exercise could seemingly go on forever. You may never get out of your living room to count the things in the rest of the house, never mind the rest of the universe. Noticing your distress, your friend pours you another glass of wine and tries to make the question a bit more tractable. It is, she says, not individual things but *kinds* of things we want to know about. This is not immediately helpful. Is she talking about square things, shaggy things, moving things, funny things, poisonous things, smelly things, dangerous things, fluid things, painful things, or what? Is our conversation right now an example of an interesting thing? *It's* in the room. Come to think of it, maybe it's only what slightly intoxicated people *believe* is an interesting thing, which could be an entirely distinct category of thing. As with the attempt to enumerate all the individual things, the list of categories or kinds seems to be endless. More wine.

What you have done in moving from individual things to categories of things is taken a step on the ladder of generalization. You could proceed step by step up this ladder, but there's no pre-ordained path upward. The next possible step, for example, might be to organize things according to the categories of the sciences. So, you will get a distinct count of what's in the room or the universe if you stick to the basic categories of physics or chemistry.

The physicist, for instance, will lump the wine glasses and the beam of light together. These are after all just different forms of organized matter, both of which obey the fundamental laws of physics in the way they appear to us. The chemist will say something analogous. And if you ask an evolutionary biologist what there is from the standpoint of her discipline she will likely focus on you, your friend, the dog and the billions of microbes crawling all over, and inside of, the three of you.

These explanations tidy things up somewhat, but your friend wants you to ascend to an even higher rung of the ladder. Why? Because you are trying to organize or enumerate *everything*, and it looks as though in answering the question the way they do the scientists have simplified through exclusion. For instance, suppose you think the physicist's answer is on the right track. It does seem to work well for wine glasses and beams of light. But what about the category of interesting things (or, if you like, 'interestingness'), including the very conversation you are currently having? That's got to be something more than, or different than, matter in motion.

The same might go for the pain in your knee and all the other psychological items you can count. Indeed, you might be inclined to extend the point beyond things like pain states, itches or moods and make a distinction at this point between all the physical things in the room on the one hand and your perceptions (or sensations) of them on the other. Now you've made some progress, because it looks like everything falls into one of just two basic categories! Take the apples in the fruit bowl on the table in front of you. Just another bit of organized matter, surely, and yet can the same be said of the experience of redness you have when you look at them? The experience is caused by the interaction between the objects and your sensory equipment. But it, the experience, seems to be nothing like those purely physical things. Let's pursue this idea with a little thought experiment.

The 17th-century German philosopher Gottfried Wilhelm Leibniz (1646–1714) asks us to imagine the brain blown up to the size of a factory so that we could travel through its maze-like structure. We'd see a lot of cool goings-on there—towering, veined and pulsating gray walls lit by a dazzling show of neuronal firings as chemical information is taken into brain cells from dendrites and transferred out of them via axons—but at no point would we encounter the smell of a beached catfish, the bright yellow of a sunflower or the sound of our favorite tenor in full throat. It's easy to see what Leibniz is getting at here. Puzzlement at how seemingly non-physical things can be reduced to purely physical happenings seems like a reasonable default position. Even so, you'd like a tour through this factory. So you book one.

Your guide, an eager neurophysiologist, is a bit miffed by your initial skepticism. But she's passionate about the brain, and convinced she can help you see the light, so the two of you jump into what looks like a golf cart to track down the precise causal pathway of, say, decayed fish-smells. She brings you to the bottom of the front of the brain, pointing out where scent molecules make their way to the olfactory mucosa, then to the olfactory bulb, from there speeding off first to the piriform cortex nuzzled just behind the bulb and then, circling back a bit, to the thalamus where smells are synthesized with information about taste. "Voila, smells explained," beams your guide. But you remain perplexed. The smell of rotting fish, you point out, is nothing like *any* of this. It has certain *qualities*: putrid, nauseating, a kind of dull sourness, indelibly linked, for you at least, to memories of the sea and to vague feelings of dread. You can't *see* any of that in this otherwise impressive tour.

The thoughts you have just entertained *might* push you to say that experiences like redness, itchiness and melancholia are a different kind of thing than those that typically cause them (like looking at apples, contact with poison ivy or a lazy afternoon). You may not have any idea exactly what such experiences are, and believe that apart from them physics is basically right. Still, you can't help thinking that at the most basic level, at the highest rung of the ladder of organization your friend is urging you to ascend, there are just two *kinds* of things: matter and what we are calling experience (really just a catch-all for the 'in your head' stuff).

In the course of a single afternoon you have discovered that you are probably a metaphysical *dualist*, someone who believes there are exactly two kinds of fundamental categories into which everything there is can be slotted. More than that, you're a Cartesian dualist. That means you think these two things—*substances* he calls them—are ontologically distinct kinds of things. Nothing we can say about one is true of the other.

We've already encountered a different brand of metaphysical dualism, with Plato and Augustine. Recall that those guys erected a basic distinction between this whole world and the one beyond it. Descartes is doing something different. He's a this-worldly dualist, insisting that all the stuff around us can be seen as ways in which the two substances—the material and non-material—are modified (have properties). If you want to divide the world up into separate substances like this, you need to decide where to put them. The sensations, or experiences, are modifications of mind-substance, but where is the mind? Is it hooked up for a time to a physical brain? How is that hookup supposed to work? How are minds and bodies, construed as distinct substances, supposed to interact causally? Tough questions.

Back at the very bottom of the ladder you notice that the wine is finished, the afternoon is nearly spent and you still have to check your work email, so you decide to call it quits. As you clean up and resume your daily routine you might think this has all been fun and heady, the way it often is when you and your friend get together, but also can't help wondering why anyone would spend more than the occasional afternoon thinking about such matters. Let's look at the life and work of someone who took them very seriously indeed, and *why* he did so.

Olympian dreams

On the night of November 10, 1619, Descartes reports having a series of dreams that, he said, changed his life. The dreams are recorded in Descartes' private notes, which he called the 'Olympica,' and so have become known as his Olympian dreams. They are strange dreams and have been variously interpreted for over 200 years. Even Freud had a go at trying to figure them out. It's the final dream in the sequence that is especially interesting (though the first contains the notable appearance of a large melon, fodder for an amazing amount of psychological speculation about Descartes over the ages). In it, Descartes discovers an encyclopedia, which he opens at random, there discovering a poem that reads, in part, "what course in life shall I pursue?" At that point, he notices a man who tells him to read a different poem beginning with the words, "Yes and No." Somehow, both poems become lost, the encyclopedia vanishes and then reappears, and the man disappears altogether.

Descartes awakens and interprets the dreams as revealing to him the "foundations of a marvelous science." The "Yes and No" poem points, he claims, to the Pythagorean notion that the mind should be directed towards discriminating truth from falsehood in human knowledge (Watson, 2002, 193; 110–111). The shadowy status of the book and its poems means that the truth should not be sought after merely in books, and not chiefly through what was then known as the study of letters. Finally, that the only other person in the dream also vanishes indicates to Descartes that he is on his own in this enterprise, that no authority figures can be fully trusted to provide answers to life's most important questions.

So what exactly was this life course and the marvelous science that drove it? Descartes' *Discourse on Method* lays it all out for us. The *Discourse* is a curious book. It is, as it title indicates, a treatise on the correct method for the sciences (a common type of book at the time), but that part of it is merely a short preface to three much larger, more purely scientific, works on geometry, optics and meteorology respectively. We'll come to Descartes' core physics in a moment, but it is worthwhile to stick with the preface for a little while. Though ostensibly an autobiography of his early intellectual development, this book contains almost no references to what was going on in the world around Descartes. This would not be especially noteworthy were it not for the fact that at the time the book was written (around 1628, though it was published, anonymously, about 10 years later), Europe was embroiled in one of the longest and bloodiest wars of its entire history to that point.

This was the Thirty Years War, with Catholic France and its Protestant allies on one side and Catholic Spain and the Holy Roman Empire on the other. The war devastated Europe. Roughly 8 million people lost their lives in it. It began in 1618, with the attempt of Ferdinand II, then king of Bohemia and future emperor of the Holy Roman Empire, to meddle in the internal affairs of other countries. It ended, fittingly enough, with the 1648 Treaty of Westphalia, which enshrined the doctrine of state sovereignty in the West. Thanks to recent scholarship, we now also know that the hardships of war played out against the backdrop of profound climatic disruption. As we saw in Chapter 3, these were the years of the Little Ice Age, when temperatures plummeted across the world and crops failed nearly everywhere. So just below the surface of war's cruder outrages, climate change–induced misery was widespread and took many forms: famine, disease, despair, political rebellion, witch burning and suicide (Parker, 2013).

In other words, the continent was a bloody mess at the time. How incongruous then to encounter Descartes' description of his time right in the middle of it all:

> At that time I was in Germany, where I had been called by the wars that are not yet ended there. While I was returning to the army from the coronation of the Emperor, the onset of winter detained me in quarters where, finding no conversation to divert me, and fortunately having no cares or passions

to trouble me, I stayed all day shut up in a stove-heated room, where I was completely free to converse with myself about my own thoughts.

(1985a, 116)

How could a person not be troubled by *any* "cares or passions" in the midst of this carnage? These tranquil plans unfold in a country, Germany, that was in the process of losing about 20% of its population to violent premature death, a figure that does not include those killed or dispossessed in various ways from the effects of climate change (Toulmin, 1992). You might wonder what was wrong with this man. His detachment looks positively pathological.

This is yet another example of how tempting it is to malign Descartes. Sometimes it can seem like everything we dislike about contemporary culture, or about our own nature, can be laid at his intellectual doorstep. He's our favorite philosopher-monster. A recent biographer, Richard Watson, compiles an amusing list of his many kinds of detractors:

I soon quit collecting pejorative uses of the word Cartesian. . . . Just pick up any book promoting religion, holism, communalism, sacralization or resacralization, any tome opposed to modern science and technology. . . . Any attack on logic, efficiency, paternalism, meritocracy, technocracy and elitism.

(2002, 19–20)

He goes on in this vein at some length, but I trust you get the picture. The complaints generally fall into one of two categories. The first is that Descartes himself was a cold and heartless rationalist, and that we should probably distrust anything he has to say because of this character flaw. The second is that this cold rationalism extends to how Descartes would have us understand the rest of the world: it is soulless and blankly mechanical, nothing more than a disenchanted and mathematized grid. Let's take these two complaints in order.

To begin, there's no reason to think Descartes was the monstrously uncaring person possibly implied in that passage about his untroubled mind. He might rather have been unaware of the full scale of the horror. He was, we should note, an officer and was never engaged in much real fighting. In any case, apart from the war context, there's evidence that his emotional economy was in proper working order, at least for a typical male of his day. For instance, by all accounts he was deeply shaken by the death of his daughter, Francine, to scarlet fever. He wrote soon afterward to a friend saying that he did not believe men should not cry at such times (Watson, 2002, 188).

Nor did Descartes believe that reason is our only important mental faculty. The very last book he wrote before he died in 1650 was a treatise on the emotions, *The Passions of the Soul*. This contains remarkably penetrating discussions of love, anger, pride, glory, shame, hatred, generosity and more. And throughout the later part of his career, he was engaged in a lengthy correspondence with princess Elizabeth of Bohemia. She found his advice about how to live well so nuanced and informed that she took to calling him the "doctor of my soul." This philosopher evidently understood more than most about how we frail, finite, social and embodied creatures actually work.

As for the disenchanted view of reality, Descartes must plead guilty. But where exactly would we be without this metaphysical innovation? Let's come back to the dreams and the foundations of a new science that Descartes claims were revealed to him in them. Descartes was in these years Europe's finest geometer as well as one its best mathematicians (he would say *the* best). The great discovery in question is what we now call analytic geometry. The basic idea behind it is that all bodies are fundamentally the same. The previous assumption had been that each science treats fundamentally different kinds of bodies. Physics deals with non-living bodies, biology with living bodies, chemistry with gaseous and fluid bodies, and so on. Descartes' dream is one of disciplinary *unification*. The properties of any of these bodies can be quantified and plotted graphically.

This is the system of Cartesian coordinates. It means that all problems in geometry can be understood mathematically and vice-versa (Watson, 2002, 115–116). This is a seminal achievement in our intellectual history. Without it, Leibniz and Newton could not have invented calculus a couple of generations later. And without calculus we would have no precise way of understanding the nature of dynamic systems. We would have only a primitive understanding, or none at all, of light, acoustics, the weather, electricity, harmonics and so on. Just think of all the particular technologies these fields of knowledge have spawned, from laser surgery to the computer modelling that has allowed us to understand the trajectories of phenomena like climate change and COVID-19. Of course, analytic geometry is also the theoretical wellspring of destructive technologies like hydraulic fracking. We will need to bear this in mind as we think through the complexities of the intellectual heritage Descartes has bequeathed us.

However, the point for now is just this: to the extent that we consider any of the technologies just enumerated (among many others, of course) a boon to humanity, we are in no position to characterize Descartes' achievement as an unmitigated disaster, as so many are wont to do. We are what we are thanks in no small part to the monomania of this supposed monster from La Haye en Touraine. Enough *kvetching* about Descartes already!

In the *Discourse*, Descartes describes his early life as the progressive encounter with three "books." The first is a set of real books, all the classics forced on him by his educators at one of Europe's finest Jesuit schools, La Flèche, in France's Loire region. The second book is "the great book of the world." Both of these so-called books were a profound disappointment to Descartes because both were marked by endless and apparently irresolvable disputation and disunity. In the first set there is hardly any agreement among experts as to proper methods and first principles. And, of course, the world at large presents a picture of deep fissure among humans. But mathematics stands above this roiling mess. Because of the "certainty and self-evidence of its reasonings" mathematics becomes, for Descartes, the methodological standard-bearer.

One metaphor that surfaces again and again in this work, and throughout Descartes' writing, is that of foundations and the structures we build on them. All of human knowledge has been built on shaky foundations, and the only remedy for this sorry state of affairs is to tear the whole thing down and begin from the ground up. This conviction is what pushes Descartes to examine the third and final book in his series, the book of himself. Descartes has arrived at the stunningly counter-intuitive conclusion that he can set the whole edifice of human knowledge on a solid foundation by closing his eyes and *meditating*.

Again, this presumption is born in perceived crisis. If you were the smartest geometer in the world, someone who has just discovered a way to unify all the sciences as well as provide us with an unprecedented degree of technological mastery over nature, you would probably feel a similar sense of frustration at the inertia of the old way of doing things.

However, Kuhn has shown us that paradigms die hard and sometimes they must be pushed over the edge. Descartes gets this. To make the world safe for science he thus realizes that he must bring everyone on board, the theologians and ecclesiastical authorities no less than the mathematicians, men of letters and even some of the great monarchs of Europe. He needs a way of presenting his insights in the most engaging manner possible. Time to sit down in the stove-heated room and enter the mind.

What's the matter?

Descartes' *magnum opus* is his *Meditations on First Philosophy*, first published in 1641. It is justly famous for the simple reason that it is, at least in its early stages, quite a bit more intellectually exciting than philosophy books usually are. And it is exciting because, through a disarmingly chatty but deceptively systematic exercise of introspection, it gets to the core of some of philosophy's most enduring puzzles. Here, we're interested in just one of these: what is matter really like?

Remember that Descartes admires mathematics because of its certainty and the way this certainty compels assent about mathematical propositions among its practitioners. With this model in place, Descartes decides that he will doubt anything that admits of even the least reason for skepticism. This, it turns out, will eliminate the entire physical world. Why? Because all the information we receive about that world comes through the senses, but sensory information is almost ridiculously unreliable. Look up into the sky on a clear night with a full moon. What do your senses tell you about the size of the moon? Look at that stick in the water, which appears bent. Is the moon really the size of a large coin? Do sticks really change shape when immersed in water?

Of course not, and in what follows all the senses are made to pay for the deceptions of a few. In the rest of this book Descartes goes on to show that even mathematical propositions are subject to doubt, since there *could* be an all-powerful evil genius deceiving us about them. Eventually, he dismisses this hypothesis by showing that God would not tolerate sharing the metaphysical stage with such a duplicitous cad. These arguments are intrinsically fascinating but are also something of a sideshow. The *Meditations* begins by casting doubt on the existence of bodies as the old science conceives them, and ends by bringing back these bodies as the new science conceives them. *That* is the frame to focus on if you want to understand Descartes' real purposes.

To get a sense of just how disruptive this transition is, consider this comment made by Descartes' meditator just after the theoretical devastation of Meditation One., i.e., just after the world has been deprived altogether of material bodies:

> So serious are the doubts into which I have been thrown as the result of yesterday's meditations that I can neither put them out of my mind nor see any way of resolving them. It feels as though I have fallen unexpectedly into a deep whirlpool which tumbles me around so that I can neither stand on the bottom nor swim up to the top.
>
> *(1985b, 16)*

This is every bit as arresting as, and strikingly similar to, Nietzsche's madman passage. Of course, it is the report of one man, alone in a room. But we can transpose the image to the age itself and obtain thereby a revealing picture of the basic existential disorientation of that historical moment. What Descartes himself experiences over the course of a single day is the experience of the Western mind over the course of the whole early-modern period, from Copernicus to Newton and beyond. This text encapsulates the malaise of a world in the process of a profound transformation. Early modernity is an intellectual whirlpool.

Let's come to the meat of the matter. What *are* bodies made up of? Nothing but those corpuscles I described earlier in the chapter, atoms coming together for a time to form iPhones, lizards, Donald Trump's hair and the Earth system. Because the behavior of these atomic particles is fully quantifiable, it can be expressed mathematically and the bodies into which they congeal can be represented on a coordinate plane. Again, bodies meet math in the form of analytic geometry. That just *is* modern physics. But to think about the world this way, you need to purify it, to get some things *out* of it.

The target of this exercise is the very broad category of sensory qualities: smells, tastes and colors, but also pains, worries, dreams, purposes and more. Descartes flushes all of this out of the physical world. Lemons are not yellow, silly. That slightly bitter juniper berry aroma is not sequestered in the two-fingers of Bombay Sapphire swirling invitingly at the bottom of your glass. Bodies do not *contain* pain or heat or roughness, only a dance of corpuscles. When these things come into contact with our sensory organs, they give rise to a panoply of experiences in us. But we must not suppose the physical world is anything *like* these experiences.

Too often, philosophers chide Descartes for having wrecked the world's mystery. But mystery is sometimes overrated. The philosophers with whom Descartes is battling about sensory qualities have an undeniably mysterious, but also patently ridiculous, theory about them. They think that bodies contain these qualities in exactly the way we sense them. They also believe these things are non-physical, and so at the heart of every physical body is a little non-physical packet of qualities or powers. This packet, or a part of it, gets triggered when the body comes into contact with us and then travels along some metaphysically special corridor into our minds. There are loads of problems with a view like this, but the biggest one is that it explains exactly nothing about why bodies affect us in the ways they do.

The 17th-century French playwright and actor Molière (1622–1673) satirized the old view by having one of his characters, a doctor trained to talk like a philosopher, try and explain to his audience how a sleeping potion works. It has the effect it does, he says, because of the 'dormitive power' it contains. In other words, it makes you sleepy because it contains the power of sleepiness. And what is the power of sleepiness? A non-physical 'thing' located somewhere inside the potion. Generalize this response to get a full appreciation of just how inane it is. Roses appear red because of the power of redness they contain. Cough medicine tastes bitter because it has a bitter power in it. The knife that rends your flesh causes you to scream in pain because it, the knife, contains pain-ness. Etc. We're well rid of these enchantments and should pat Descartes and his philosophical confrères on the back for expelling them from the physical world.

Modern science is genuinely revolutionary because it applies these insights to reality in such a *comprehensive* way. As we have seen, this is nicely exemplified in Newton's unification of the sublunary and superlunary spheres. But it is even more applicable to the new epoch. In a striking recent analysis, the philosopher Christopher J. Preston refers to this new time as the Synthetic Age. In this Age, he argues, design "will reach deeply into the Earth's metabolism" (2018, xviii). Preston does not offer an unambiguous cheer in favor of these developments. Rather, his aim is to show us (a) in what sense they are actually being realized in our current technologies and (b) to emphasize the need for sustained social and political dialogue on how they proceed.

I've already said something about that second point (in Chapter 2) and will return to it later in this chapter and in Chapter 9. So let's focus briefly on the first point by isolating three broad technological developments that will define life in the new epoch for decades to come. I've selected them to emphasize the comprehensiveness of the Cartesian innovation, the point that every aspect of our spatial reality is up for grabs. The three areas encompass distinct, though interconnected, levels: the atmosphere, the planet's makeup of non-human species and our own bodies.

First, we are now actively looking for a way to design the chemical makeup of the atmosphere. This is geoengineering. Its most widely discussed variant is a form of solar radiation management known as sulphate injection. The idea is to spray sulfur dioxide particulates into the atmosphere, where they will bounce incoming sunlight back to space, thereby preventing the sunlight's thermal energy from being absorbed by the Earth. We got the idea from the effects of sulfur-dioxide-spewing volcanic eruptions, the biggest of which have cooled the planet measurably for a short time afterwards. Applied intentionally, something similar might work, would be relatively cheap and is probably scalable.

There are other variants of geoengineering, not all of which have these qualities but that might be less risky. We need to have a conversation about who benefits from geoengineering, who controls the global thermostat and how dangerously deep our ignorance is of the ecosystemic effects of these technologies. Even if we decide that sulphate injection is too risky we have not thereby absolved ourselves of the responsibility of engineering the chemical composition of the atmosphere *in some way*. We're already doing this on a massive scale with our more or less unconstrained carbon emissions.

The second cluster of technologies will be aimed at genetically modifying or relocating existing species. The purpose of the first strategy is to protect species against extreme changes in their habitats for which they are not sufficiently adapted. We are already doing this with plant species, to make them more resistant to drought, but the technologies can be extended to animal species too. The goal is to modify the animal at the genetic level so that descendants of the population can withstand various climatic alterations. If this is not feasible, then the second strategy—relocation—might kick in.

Again, we're doing this already on a fairly significant scale, especially in Europe where the 'rewilding' movement is well developed, but the efforts are likely to become much more aggressive as species begin crashing more quickly than they are now (Monbiot, 2014). Now, genetic manipulation and relocation obviously involve distinct methods of technological intervention into the non-human sphere. But both represent an effort to *design* that sphere and are therefore caught up in questions about what is valuable and why, what can be saved and what cannot, and so on. The main problem with all these techniques is the same as it is with geoengineering. Ecosystems are fantastically complex systems. In many cases we don't know what the knock-on effects of our interventions will be. We could end up causing even more destruction than climate change would.

Finally, there are technologies of the human body. These come in two broad varieties. The first concerns the possibility of genetically enhancing humans so that they are better able to cope with life on a hotter planet. That would have obvious adaptational benefits for our descendants. The second is pharmacological moral enhancement. Here, the idea is that people would take drugs that enhance their altruism and sense of justice. If our lacking these two motivations is what has gotten us into the mess we are in, then we need to have a conversation about how to fix this and technological interventions like these are becoming a part of these conversations. You might think this is outrageous, but, again, it is being taken seriously, and this fact alone is a good indication of the novelty of our predicament (Persson and Savulescu, 2014).

These two forms of technological personal enhancement, the genetic and the moral, might be beneficial for humanity. But the risks they bring are clearly grave. Both forms of enhancement raise deep concerns about how 'normality' is defined and the way this definition can discriminate against those who fall outside it (Hall, 2016). Humanity's sporadic flirtations with eugenics might, all by itself, be a reason to declare a moratorium on research into any of these technologies. Again, the point here is to highlight the fact that these conversations are happening and that grasping the full complexity of our conjuncture requires appreciating this.

Now, you might be thinking 'why not simply reduce our emissions of greenhouse gasses rather than engaging in all this risky engineering?' Well, even if you believe we should engineer aggressively it is misguided to think that we should *also* continue to super-charge the planet's energy system by pumping more and more CO_2 into the atmosphere. Why make the building task more onerous than it has to be? This puts paid to a broad worry about technology that fits under the heading of 'moral hazard.' A moral hazard is created when people believe they are protected against a hazard, and so behave more recklessly with respect to that hazard. For instance, other things being equal, you will drive less cautiously if you believe you are insured for potential damages to your car resulting from a crash.

But building in the age of climate crisis isn't like this. In the coming decades there is no good reason to believe that any technological measure we take is going to protect us in a fully reliable or permanent way. It would be the height of folly to suppose that we can safely dig up and burn the world's remaining coal reserves, for instance, just because we think we've got the technologies to protect ourselves fully from the climatic effects of this decision. We don't and there's no reason to believe we will in time to avert the disaster that would befall us were we to refuse to decarbonize the global economy as swiftly as possible.

Conclusion

I am not blindly optimistic about our chances of designing the technosphere in a way that will save us. A better way to think about what I'm up to in this book is that I'm taking a realistic view of where we actually are. In very broad strokes only two courses of action are open to us. We can either refuse or accept the invitation to design the technosphere intentionally and ethically. Thanks in large part to anthropogenic climate change, comprehensive, metabolic design is already here. Taking the first course—refusal—will therefore not grant us access to, or allow us somehow to recover, a non-technosphere. It will only allow the accidental features of the megastructure to proliferate—making it less beautiful, less heterogeneous, less sustainable, apt to generate benefits and burdens that are less equitably distributed and so on.

Anyway, amid all the prophecies of doom besieging us these days, is a shot of optimism such a bad thing? I talk quite often to engineers about climate change.

I like doing this because they are usually pretty upbeat people, and I really *need* the occasional infusion of optimism about this issue. But generally speaking they are also, I must say, quite naïve about the dimension of this issue that has to do with values. To return to a distinction discussed in Chapter 3, with a few exceptions professional engineers tend to assume that technological progress always leads to moral progress. The latter will somehow take care of itself so long as all the right tools are in place. As we've seen, there's no good reason to think this way. We have to fight to make sure it happens.

Descartes has given the West a systematic metaphysical dualism between two kinds of substance: minds and bodies. I've argued that this dualism is a response to a sense of intellectual crisis that was also an existential crisis. The only way to make the world safe for science's high moral purpose was to rid it entirely of anything that could not fit into the system of mathematically representable coordinates. This is the process of purification that has given us the technosphere. For urban geographer Stephen Graham the technosphere is,

> a kind of carpet covering large areas, on which the furniture of the human world (its buildings, bridges, monuments, pylons, oil-rigs, telegraph poles, roads, railway viaducts, cities, shantytowns, parks, airports) stands and is supported, and into which it will eventually crumble. Deep-layered in places, threadbare and patchy in others, this carpet of near-global extent provides the surface on which people carry out their lives. Like a carpet, it is so well-used it is taken almost totally for granted.
>
> *(Graham, 2016, 292)*

It's this taken-for-granted quality of the technosphere I want us to resist. One way to do so is by pressing the worry that the carpet is too homogeneous and that Descartes himself gives us no theoretical resources for repairing this defect. Is there a way of embracing *both* an expanding technosphere *and* the rich heterogeneity of the non-human? Indeed there is.

References

Bacon, F. (1905). "Novum Organon." In *Works*. London: Routledge.

Descartes, R. (1985a). "Discourse on the Method." In *The Philosophical Writings of Descartes*. Cambridge: Cambridge University Press, volume I.

———. (1985b). "Meditations on First Philosophy." In *The Philosophical Writings of Descartes*. Cambridge: Cambridge University Press, volume II.

Graham, S. (2016). *Vertical: The City from Satellites to Bunkers*. London: Verso.

Hall, M. (2016). *The Bioethics of Enhancement: Transhumanism, Disability and Biopolitics*. Lanham: Lexington Books.

Kuhn, T. (1970). *The Structure of Scientific Revolutions*. Chicago: The University of Chicago Press.

Monbiot, G. (2014). *Feral: Rewilding the Land, the Sea and Human Life*. New York: Penguin.

Parker, G. (2013). *Global Crisis: Climate Change and Catastrophe in the 17th-Century*. New Haven: Yale University Press.

Persson, I., and Savulescu, J. (2014). *Unfit for the Future: The Need for Moral Enhancement.* Oxford: Oxford University Press.

Preston, C.J. (2018). *The Synthetic Age: Outdesigning Evolution, Resurrecting Species and Reengineering our World.* Cambridge, MA: The MIT Press.

Toulmin, S. (1992). *Cosmopolis: The Hidden Agenda of Modernity.* Chicago: University of Chicago Press.

Watson, R. (2002). *Cogito Ergo Sum: The Life of René Descartes.* Boston: Godine.

Zalasiewicz, J., et al. (November 28, 2016). "Scale and Diversity of the Physical Technosphere: A Geological Perspective." *Anthropocene Review* 4(1), 9–22.

7

SPINOZA: DIVERSITY IN UNITY

I'm pretty lazy, truth be told. Years ago, I decided to give up on weeding my garden and lawn. Why bother wasting your Sundays doing *that* when there's always a game to watch? And why are we so obsessed with this activity anyway, with eliminating the first trace of plants that look a little out of place and are somewhat difficult to control?

The whole weeding business seems a bit pathological to me, frankly, but I have noticed that my indolence on this front has made me something of a neighborhood pariah. The woman down the street won't even allow her preciously prancing Alaskan Malamute to *defecate* in my front yard, though for some reason it loves the spot. As it wiggles down every morning into that tell-tale crouch just above my beautiful dandelions, she tugs it firmly along, muttering through gritted teeth, "Not *here*, Cleo!" On those rare occasions when I decide to cut the grass, my next-door neighbor won't loan me his mower for fear that my weed-seeds will hitch a ride on the machine's undercarriage, migrate to his pristine patch of green and pollute the whole damn thing. Heaven forfend.

Faced with such outrageous reactions, I've dug in my heels. I've decided to find a way of turning my vice—laziness—into a virtue, or at least finding a philosophical justification for it. And lo and behold, it turns out that cultivating weeds is a good way of fighting climate change. Take Australia, in many ways ground-zero for climate change. This place has suffered devastating bushfires, drought, dust storms and soaring temperatures. In January of 2019, Birdsville, Queensland experienced 10 straight days of 45°C, but extremes like this have hit the entire country. Temperature records are shattered again and again. Australia, of course, is also home to some of the world's most magnificent coral reefs, extraordinary repositories of biological diversity which are being bleached to death by rising ocean temperatures.

Enter Peter Andrews, an Australian farmer who refers to his country as "the laboratory of the world when it comes to adapting to the weather." Andrews has made it his mission to restore drought-ravaged landscapes to a condition bursting with carbon-sucking flora. To this end, he has developed a technique known as natural sequence farming. It's a complex process, but a key element of it involves allowing weeds to flourish with utter abandon. Just east of Canberra, in a place called Mulloon Creek, there's now a small pilot sight testing out the idea. The results have been so positive that in 2016 the UN commended the operation as one of only a few genuinely sustainable farming sites in the world. The main result of natural sequence farming is to increase water flow and raise the water table over the affected area. This can make it an ideal farming technique for areas hit hard by climate-induced extreme drying (Kenyon, 2019).

If you're going to embrace a technique like this, you need to accept some messiness in your world. For example, natural sequencing is as happy to work with invasive species as endemic ones, the sort of attitude that drives conservationist purists mad. But if it works to stabilize soil, retain water and sequester carbon, who cares what the plant's point of origin is? Quite apart from the beneficial climatic effects of this approach to farming, what I like most about it is its tolerance of spontaneity and diversity in the built landscape. Being in the Anthropocene evidently does not entail making everything over in the same way. It is not a call to homogeneous building, as some have supposed. That, at least, is what I'm going to try and convince you of in this chapter.

All of this is a carry-over from the previous chapter's focus on purification. By the end of the *Meditations* Descartes believes that all the pieces are in place to reunify the previously broken world. In fact, one way to look at the radical doubt with which that book begins is that Descartes decides to break the world even more completely than it had been before. The whole tottering structure needs to be brought crashing down, so pervasive is the rot. As we have seen, after that exercise Descartes bewails the angst-inducing topsy-turviness of the world made—or unmade—by his own stove-heated reflections. By the end of the book, of course, the world is put back together again. Everything in its place. Home. But only for the scientist and for the rest of us inasmuch as we are the beneficiaries of science's mastery of nature.

The big question is this: can we accept Descartes' purification of the material world—the one that produced the technosphere—while also celebrating and helping to enhance its rich and mostly non-human diversity? Guided by that question, we'll explore the thinking of one of the most brilliant and iconoclastic philosophers the world has ever known, Baruch Spinoza. As we'll see, Spinoza answers the question in the affirmative. But in my view he could not have come up with the answer absent his own crisis, which is simultaneously a crisis for his people. We begin there.

The Sephardim

Spinoza (1632–1677) was a member of Amsterdam's Jewish community. A probably apocryphal, but very romantic, story has it that this community originated from a Spanish ship carrying Jewish refugees from Portugal sometime between

1593 and 1597. At the time, Britain and Spain were at war, and the ship was waylaid by the British navy. Aboard were one Maria Nuñes and her relatives. It is said that Maria was so beautiful that the captain of the British ship, a duke, fell in love with her and asked for her hand in marriage.

She refused the offer and was eventually brought before Queen Elizabeth. The queen was equally charmed by Maria, who quickly became the darling of court society, with many opportunities for marriage and general social advancement presented to her. She refused all of this so that she might finish her journey. Her destination was the Low Countries, to which she eventually made her way. She married her cousin in Amsterdam, and their household established the so-called Sephardic community in that city. Spinoza would be born into it some 35 years later (Nadler, 1999, 5–6).

Who were the Sephardic Jews? To answer this question we need to go back about a hundred years prior to Maria's voyage, to Spain's expulsion of its Jews on March 31, 1492. Here is part of the proclamation issued by Ferdinand and Isabella, who had become King and Queen after the union of Aragon and Castile in 1479:

> We have decided that no further opportunities should be given for further damage to our holy faith. . . . Thus, we hereby order the expulsion of all Jews . . . who live in our kingdom. . . . Jews shall not be permitted in any manner whatsoever to be present in any of our kingdoms and in any of the areas in our possession.
>
> *(quoted in Nadler, 1999, 3)*

Previous to this edict, Spain's Jewish community had already been living a precarious double existence. Their religious freedoms had been steadily eroded since the early years of the 15th-century. Violent attacks on them by the larger population were frequent. Just then, the Spanish Inquisition was increasing its surveillance of subjects and the general suppression of heresy. Many Jews were forced to convert to Christianity. Some of them became sincere Christians, but many others did not. Instead, they wore the mask of the dutiful Christian subject while maintaining Jewish religious rituals underground. They were therefore known as *conversos*. We would say now that they 'passed' as Christians.

Segregation, hiding, expulsion, exile, conversion, passing. These are the terms defining Jewish life in this part of Europe at that time. Needless to say, this is an almost textbook example of what it is like to be homeless, even where one is ostensibly at home. After the official exile from Spain, things got even worse. Many Jews made their way to Portugal, which was relatively tolerant of their presence for a time. But this too did not last. In 1497 Portugal's ruler, Manuel, ordered the forced conversion of all Jews and the baptism of all Jewish children. After the union of Spain and Portugal in 1580, the Inquisition moved into Portugal with a vengeance. Thus began the steady migration of Jews from this area to northern Europe. By 1609 they were firmly established in Amsterdam and other Dutch cities (among many other places in Europe, Northern Africa and even the Americas) (Nadler, 1999, 4). These were the Sephardim (literally, 'the Jews of Spain').

The Low Countries were an especially attractive destination for these people because of the relative toleration of the Dutch people and government to religious diversity. But to overstate this point would be to misunderstand a potent source of the Jewish community's lingering sense of insecurity. Dutch society itself was somewhat torn by religious strife at the time. The politically dominant Calvinists were wary of the Reformed dissenters in their own midst and this made them sensitive to all religious practices that failed to conform to strictly orthodox standards.

So they tolerated the Jews, whose commercial prowess they valued, but this attitude of forbearance walked a knife-edge, as did the fortunes of the Jewish community itself. Indeed, not until 1657 did the Jews of Amsterdam gain status as citizens of the republic, acquiring thereby some measure of official protection from arbitrary harassment. Even then, they did not enjoy unfettered freedom. There were severe punishments for anyone found attempting to convert a Christian to Judaism, for example (Nadler, 1999, 12–14). This is the context in which we must set the life and fate of our philosopher.

Think of what the Dutch Jews had managed to achieve in such a short time. Spinoza's recent biographer, the philosopher Steven Nadler, describes Amsterdam's Jewish community as "a rich and cosmopolitan but distinctly Jewish culture" (1999, 26). So: a thriving community, but one whose prosperity might be snatched away at any moment. A community at the mercy of larger social, political and cultural forces. And, just as important, a community with a living memory of what can happen when the political winds shift. Because the forces of Calvinist orthodoxy could turn on them in a heartbeat, the leaders of Amsterdam's Jewish community bade their flock to keep a low profile.

This was not in the DNA of Baruch Spinoza. From the standpoint of the religious authorities, the main problem with his thinking was that his metaphysics overturned nearly all the fundamental assumptions and precepts of monotheism, including Calvinism and Judaism. We'll come to that in a moment. Also problematic was his political philosophy, which gets far less attention among current philosophers than his metaphysics. Spinoza was a fierce defender of religious toleration and individual freedom. Government, as far as he was concerned, was there to protect the speech and actions of its people, not to suppress them. This was entirely too *laissez-faire* for the more politically circumspect Dutch authorities, and by extension the Jewish community leaders as well (Nadler, 1999, 148). But Spinoza did not say what he did because he enjoyed ruffling the establishment's feathers. He was no glib contrarian.

Rather, he was wholly focused on pursuing the truth wherever it led, and the political and social consequences of his findings were irrelevant to him. He never seemed to waver in this belief. And so, after repeated warnings to desist from his dangerous thoughts, on July 27, 1656, the Jewish leaders made the following proclamation at the Houtgracht synagogue:

> By the decree of the angels and by the command of the holy men, we excommunicate, expel, curse and damn Baruch de Espinoza, with the consent of God, blessed be He, and with the consent of the entire holy congregation. . . . The Lord will not spare him, but then the anger of the Lord and His jealousy shall smoke against that man, and all the curses that are written in this book shall lie upon him, and the Lord shall blot out his name from under heaven.
>
> *(Nadler, 1999, 120)*

Now, it is important to point out that this directive, known as a *kherem*—meaning, 'set apart,' 'destroyed' or 'cursed'—was issued well in advance of Spinoza actually publishing any of his ideas. He had, however, evidently felt little compunction about kibbitzing philosophically with friends, and because the Amsterdam Jewish community was tightly-knit word of his heterodox opinions spread quickly (Goldstein, 2006, 30). When it found its way back to the rabbis all hell broke loose.

I always make a point of reading the *kherem* out to my Modern Philosophy students. Many are visibly shaken by the edict's ferocity. And yet when the pre-history is explained the surprise and horror diminish somewhat. Here is a community, the Sephardic Jews, literally running for its life for 140 years. Finally, it manages not only to establish itself in a place where it can put down some roots but actually *thrive* there, culturally and economically. But this relative peace and security comes with terms. By your own lights these terms are not draconian, so why not do whatever is necessary to live by them?

And then into your midst bursts a rare genius with no respect for authority and a take-no-prisoners approach to the pursuit of truth. Put yourself in the place of those elders. Would you not react with similar fury? The *kherem* is an ancient coercive measure designed to enforce the casting away of something deemed unholy. The thing to be separated or removed might be considered so potentially polluting that it must not only be placed to the side but prevented from ever coming into contact with the unpolluted. Quarantined. Spinoza's *kherem* also forbade any other member of the community from communicating with him in any way, from coming into his proximity or from reading any of his writings (Nadler, 1999, 121). He was effectively to be destroyed as far as his future potential impact on Amsterdam's Jewish community was concerned.

And again, just think of how explosive Spinoza's musings must have been for the community's leaders to want to squelch them so thoroughly before they even saw the black of print. The political ideas alone could never have resulted in this sort of action against him. To understand it fully, we must explain the metaphysics, which must have been simmering away in Spinoza's mind for years.

Deus sive natura

As a young man, Spinoza was Europe's most astute interpreter of Descartes' philosophy. Recall that one of Descartes' most important concepts is that of substance. A substance is reality's most basic constituent, so that in asking the question of how many things there are in the universe, at the most basic ontological level, we are effectively asking how many substances there are. The metaphysical dualist says that there are just two, in Descartes' case thought and matter. Where to go from here? Well, what Spinoza also finds in Descartes is the claim that one of the chief characteristics of a substance is its *independence* from other things. This independence, in turn, makes substances just a little more *real* than non-substances.

This is all quite abstract, so let's try and ground it a little. What's the connection between independence and being real? Isn't everything that *is*, equally real? In a sense, sure. But these philosophers, remember, are operating in an intellectual context in which the idea of God is ever-present, and all of them were forced to say something about what makes this Being different from, but also related to, everything else there is. From within that worldview it is not at all bizarre to make the link between independence and reality.

Just compare God and a rock, the Thing at the top with one of the things at, or near, the bottom of the Great Chain of Being. Rocks go in and out of existence all the time. The degree of permeability they have will affect the rate at which they succumb to processes like weathering—compare diamonds to limestone, for instance—but they all break down sooner or later. More importantly, they remain solid for as long as they do only with the help of *other* things. Diamonds can form only deep in the Earth's mantle where the temperatures and pressures are really high, and these areas can be found in only a limited number of places on the planet. In the absence of this precise set of surrounding material conditions, diamonds will never form. They therefore need *help* to develop. This is true of every physical thing you can think of.

God, by contrast, does not depend on anything other than Itself for Its existence. One way in which this idea has been articulated over the ages is to say that God is self-caused, *causa-sui* in the Latin. Aristotle says that God is the Unmoved Mover. To be a mover is to be capable of being the cause of the movement of other things, but to be unmoved is to have nothing outside yourself that is capable of doing the same thing to you. That's the independence bit. Radically independent things like God are the most real just because they cannot be destroyed and are not created. Whatever else you think about an Unmoved Mover, if such a thing exists in the first place, you must surely grant that it is thus really *real*!

Alright, now just add the final thought: that independence conceived of in this manner might be something that comes in *degrees*. A diamond has more of it than a chunk of limestone because it's comparatively easy to destroy the limestone. If it's exposed on a rock face, a few decades of hard rain should do the trick. Or suppose you believe that humans have an immaterial soul, placed in us by our creator at the moment of conception. Since only God is capable of creating and destroying souls that makes us *relatively* independent, more independent than a diamond for sure.

Have you ever marvelled at stories of religious martyrs, seemingly going down—or up in flames—without a crack in their self-confidence? That's because they believe that the flames of the auto-da-fé can't touch what they really are, that granitic center that only their God can destroy. So long as they are on the up-and-up with the man upstairs, nothing truly destructive can happen to them down here.

With that basic metaphysical picture in place let's come back to the contest between Descartes and Spinoza. Think of the relation between a mental substance and the thoughts it has. Right now, I'm thinking about the words pouring slowly onto the screen in front of me as I type. Those particular thoughts obviously *depend on* the mind in which they inhere. Thoughts don't just float free of particular minds, like fireflies lighting up the night sky. Even if I find it hard to say exactly what my mind is—is it just my brain, or something else?—I'm pretty confident that the thoughts I'm having need some such place to settle down in.

But the mind itself does not depend on those thoughts. Right now, I might have entirely different thoughts or none at all. Think of the flow of images in ads on TV or the Internet. Advertisers know how to keep your attention riveted on the screen by throwing a truly dizzying number of images at you, many of them well under a second in length. That's an extreme example of what happens all the time with perception. As you move your eyes across a scene in front of you, the details change from one moment to the next, a display made even more complex by the addition to the tableau of sounds, smells and even tactile qualities (imagine a wind suddenly springing up). And it's not just perceptions. While looking at a cow off in the distance, you might be struck by the desire to eat a steak for dinner only to realize a few moments later that you recently decided to help save the planet by cutting beef out of your diet.

Again, the point is that I'd still have the mind I do, still consider myself as myself, in the absence of any of these *particular* thoughts. So my mind does not depend on them the way they clearly depend on it. The substance that is my mind is therefore more independent than the thoughts that modify it at any given time.

However, if we are going to talk about substance in this way, then Descartes must, it seems, add a third item to his list of substances. There's mind-stuff and material-stuff, to be sure, but what about God? After all, as we have just seen, God is usually conceived of as the most independent thing there can be, so shouldn't we call It a substance too? Descartes admits as much, pointing out in his *Principles of Philosophy* that "there is only one substance which can be understood to depend on no other thing whatsoever, namely God" (1985, 210). Spinoza will take Descartes at his word here, and the sky will fall as a result.

Spinoza begins his masterpiece, the *Ethics*, with a definition of substance straight out of Descartes. Substance is that which "exists in itself and is conceived through itself" (Spinoza, 1994, 85). Though abstract in the extreme, it's really a simple idea. Here's this thing, substance, about which we must say two things.

On the one hand, its existence cannot be attributed to anything other than itself. It was not *born* of anything. And on the other hand anything that happens to it, any change it undergoes, cannot be explained by reference to anything other than itself. If you find this perplexing, just think of how it applies to the traditional idea of God. This Being was not created, and It does not suffer changes emanating from outside of It. God thus fits the definition of a substance perfectly. Now consider three momentous implications of these simple claims.

The first is that no matter how many of these substances there are they cannot interact causally with one another. For if they do interact, then what goes on in one can only be fully explained by reference to the other, thus violating the definition of substance just enunciated. This is a dagger aimed directly at Descartes, who thought that minds interacted causally with bodies but never managed to explain how this might work. Let's linger on this problem for a moment.

Right now, I'm working my way through a glass of the vegetable juice I choke down every day in a pathetic effort to eke out a few more years of life. It's called a 'smoothie' but large quantities of kale and collard greens render it more furry than smooth, so I call it a 'furrie.' To speak with Descartes, the tactile feel of this vile farrago on my tongue (its 'furriness') is a modification of my mental substance. It's a way my mind is qualified at this moment. But my mind was caused to have just this quality by certain goings-on in physical substance, namely the specific arrangement of corpuscles that is the furrie's true nature. So I can explain what's happening in one substance only by reference to what is happening in another. That can't happen, according to Spinoza. Therefore no matter how many substances there are, they can never communicate with one another.

This, it turns out, isn't so unnerving because in fact there can only be one substance. This is the second implication of the definition of substance. For suppose there are two substances. Then there must be a place where one stops and the other begins. There will be a boundary between the two. But a boundary partly defines the things on either side of it. My furrie does not spill out all over the desk because it is bound into a specific, cylindrical shape by its glass container. Suppose I want to know why the liquid is shaped just that way. This is a question about the liquid but notice that it can only be answered by pointing to something that is not the liquid, namely the container.

None of this should give you metaphysical fits when it comes to understanding the nature of medium-sized physical objects, the ordinary furniture of the material world. But things are different with substance. This is because there cannot be *anything* outside of substance without violating its definition. It can tolerate *no* boundaries to itself. It is therefore literally unlimited. The same considerations apply to its temporal aspect. If there's a time when it is not, then this is a kind of limit or boundary to it. Not allowed. So it's eternal as well as infinite. Obviously, this means there can only be one such thing. Goodbye dualism, hello *monism* (literally, 'one-ism').

But it's the third implication of the definition of substance that must really have agitated Spinoza's co-religionists. For if there's just one substance, and it is literally infinite and eternal, then there's no sky-God in a realm apart. God, in fact, is *nothing but* the totality of stuff existing infinitely in space and eternally in time. There could not have been a time prior to the existence of all this stuff, just as there can be no place where it is not. For these reasons, there was no act of Creation, because to posit one is again to assume that God is bounded, in this case by Its own Creation. With this, Spinoza has dragged the personal God of Isaac and Abraham, the literal Father and Lord of all, from His transcendent throne and scattered Him among all the stuff of the world. God is everywhere but also, in a sense, nowhere. Not the sort of thing you'd bother praying to, that's for sure.

In his inimitably precise way of putting it, Spinoza calls this *Deus sive natura*, God or nature. Again, these are maturely worked-out propositions laid out in the painstaking geometric order of the *Ethics*, not the cavalier public musings of a precocious 20-something. But supposing the young Spinoza was broadcasting even a relatively inchoate version of them among his friends, can you appreciate why the rabbis became so alarmed and infuriated?

Complexity is all

The monist will deny that there are any distinctions in reality at the level of being or substance. Again, think of this in terms of the Great Chain of Being. That trope informs us of hard distinctions at this level. Birds and bees are metaphysically distinct kinds of things from trees and rocks, to say nothing of humans and angels. This means that nothing in nature could either have emerged from or evolved beyond a different level than the one it finds itself in. For this reason, the notion is fundamentally hostile to evolutionary accounts of species development. The claim that everything living has evolved over unimaginably long time spans from a primordial chemical soup is unthinkable from the standpoint of the Great Chain of Being. So long as they do not believe in God or some other supersensible entity, all those who endorse evolutionary accounts of life are therefore monists. Darwin is a monist.

Spinoza's metaphysics is not evolutionary, in any sense of that word. This is an aspect of his thinking we will want to correct, beginning in the next chapter. For now, we need to dig a little deeper into his brand of monism. One way to do so is to ask Descartes' question: is there any meaningful distinction between matter and mind? Remember how natural it felt to think there *was* in the course of your afternoon chat with your friend, back in Chapter 6? The sensation of red in your head when you look at an apple on the table just does not seem like the same kind of thing as the purely physical thing—the apple—that caused it.

In the previous chapter, we were focused above all on the way Descartes purifies the world by driving sensations out of it. But we also broached the question about where to put all the banished qualities, the sensations and pains and emotions. The answer, recall, is: in the mind. And because only a non-physical thing can support or house non-physical modifications, the mind must be a non-physical thing or substance. What does Spinoza say about all this?

Despite being a fierce critic of Descartes, Spinoza is so immersed in the Cartesian worldview that he cannot help but provide a Cartesian-sounding analysis of the mind-body problem. Reality, he says, is fundamentally *both* material *and* thinking. Wait a minute! Isn't this basically what Descartes says? No. Descartes says that everything, apart from humans, is *either* material *or* thinking. God is pure thinking, matter—including all non-human animals—is entirely unthinking, and humans are a complicated mash-up of the two. Spinoza says that everything, including humans, is both thinking and material. The difference is absolutely critical. Why does Spinoza hold this view? Not surprisingly, it all comes back to the definition of substance. For if mind were in one place and matter in another, and both were substances, then each would limit the other. This would violate the definition of substance.

It follows that we can look at all of reality as thinking or all of it as material. But it's the same thing we are looking at in each case. For a helpful analogy, think of the famous duck-rabbit image. The image shows a duck and a rabbit, depending on how we look at it. It's both and neither (that is, neither *exclusively*). It's the same thing with Spinozan substance, all the stuff there is. Humans therefore are not *substantially* distinct from rocks or zebras. There's thought in all of it and matter in all of it too. But surely there is some important difference among these things?

The only difference, for Spinoza, is one of bodily complexity, which is reflected in the mental complexity of the thing. Physically, rocks are pretty simple, zebras much more complex and humans more complex yet. This complexity is then reflected in the thinking of each of these things. In fact, it makes sense to say that human thinking is the most sophisticated form of thinking we know of, but Spinoza will say that this can be explained fully by the corresponding complexity of the human body (which, importantly, includes the brain, the most complex physical object there is). That's just right, isn't it?

Or is it? Spinoza has apparently committed himself to some pretty wild claims here. Are we going to let stand that bit about thinking rocks? It's required by the metaphysics, but can we make sense of it? This is not an easy task, but perhaps we can begin to understand the claim if we construe thinking as the ability to communicate, in some sense of that concept. Geologists routinely *read* rock formations to inform us about past ages of the Earth. So what are geologists if not highly trained rock-whisperers? Their training consists in interpreting what's already *there* in the rocks, as though the messages were just waiting for someone to come along and read them. The point is even easier to make with trees, which new research suggests are engaged in constant and sophisticated processes of communication among their arboreal brethren as well as the fungal world (Wollheben, 2016). It's a trippy idea, which gets an evocative treatment in Richard Powers' 2018 novel, *The Overstory*.

And by now we should have no problem assenting to the notion that non-human animals can communicate. Ants, for example, do so with each other through pheromone-scented trails while dogs, chimps and dolphins do so both with each other and with us. Much work needs to be done to fill out these accounts, of course, but current scientific understanding of them gives us no reason to dismiss Spinoza's ideas out of hand. Maybe, in some sense we cannot yet comprehend fully, there really is mind *everywhere*, dizzyingly complex networks of communication going on within the systems that make up the biosphere. We'll return to this theme in the next section.

What about humans? For Spinoza, as for Descartes, humans are composite things: we think and are material. At first blush, comparing the two philosophers on this theme can make Spinoza look less bold than we expect him to be. But even here appearances are misleading. For we are not minds *in* bodies as Descartes supposes. The Cartesian supposition is rooted firmly in our status as the kind of things we are created to be. We cannot change. Because Spinoza thinks our minds are nothing but reflections of the complexity of our bodies, and he rejects the account of divine Creation that grounds Descartes' views (because there can be no Creator God), he is able to say that we *can* evolve into something beyond our current state.

So his brand of monism allows for two intriguing possibilities. The first is the transformation of human nature through various forms of technological enhancement. The second is the creation of machines that outstrip us in intelligence because they are more physically complex than us. In other words, Spinoza has cleared a metaphysical path for cyborgs and artificial intelligence. Any particular manifestation of complexity is merely temporary, and there's no reason in principle for hyper-complexity, as well as the specific level of intelligence that necessarily accompanies it, to be confined to carbon-based systems like ours. It could just as well be a property of silicon-based systems, or of literally any other kind of system. Spinoza is thus way ahead of his time, further ahead than even he realized.

There's another point to make here, one that distinguishes Spinoza from all the philosophers covered in the three previous chapters. Each of those philosophers—Plato, Augustine and Descartes—responds to crisis by splitting reality in a fundamental way. Plato and Augustine construct an otherworldly alternate to this world, while Descartes sets up two worlds within this one. The distinction between otherworldly and this-worldly dualism is important, but the urge is the same in each case. It's that urge that Spinoza rejects, and this is what makes him a truly revolutionary figure in our history. I want to elaborate this point by circling back to the story we have been telling about Spinoza's life and that of his community.

Persecution and exile have been sadly enduring features of Jewish life throughout history, of course. The story of the Sephardim is one such sorry case among many others, before and since. Exile is the evil twin of purification. When we introduced Descartes in the previous chapter we touched on this briefly, the way for instance concerns about racial or ethnic purity are almost always backed up by calls to *get rid of* a certain group of people. In my view, Spinoza is the most thorough opponent of exile the world has ever known. Why? It's customary for decent people to deplore any attempts at racial or ethnic purification, but how many of us take this a step or two further and insist that fundamental division must be rejected *everywhere* in reality?

This is why it is so important to point to Spinoza's ethnic identity in discussing his metaphysics. It is one thing for anyone—me, for example—to say 'All is one, I'm a monist,' quite another for a Jewish philosopher to say this. And not just say it, but to demonstrate that monism is an indisputable metaphysical doctrine, the true upshot of the modern scientific and philosophical revolution. In putting forward this doctrine Spinoza is not recommending a merely incremental revision to the tradition, albeit a slightly cheeky one. He is rather undermining the dualistic bent of the whole history of metaphysics in a manner that also sweeps up all monotheistic theology. Because this theology partakes of the dualism, this entails challenging its fundamental assumptions about the nature of God.

It's an extraordinary vision from someone whose identity is marked so thoroughly by the scars of persecution and exile. Remember, though he was born in Amsterdam Spinoza is not very far removed in time from the Spanish and Portuguese expulsions. Those wounds were still fresh in the Amsterdam community, and this, I've been suggesting, explains the caution of the Jewish leaders. But let's speculate a little about how Spinoza must have taken their directive. As it happens, after his own expulsion he didn't say much more out loud about *Deus sive natura*.

But isn't it possible he took the *kherem* precisely as a spur to writing everything down systematically, possibly for the whole world eventually to see? The *kherem* is yet another officially sanctioned attempt at social purification. Spinoza might have believed that unless you could show that there are no fundamental divisions in the world—neither among humans nor between us and the rest of nature—demagogues would always rise up seeking to appease the latest seekers of purity. But if you can prove all is One you have thereby cancelled the very possibility of fundamental separation. You have banished banishment, exiled exile.

What a sad irony that the philosopher who articulated this powerful idea would be confined by his own people to a life of internal exile. In 1670, Spinoza moved to The Hague, where he worked on his philosophy for the rest of his life. He made ends meet by grinding lenses, an occupation that eventually killed him just seven years later. There's something poetic about that death, after all. Convex lenses are thicker in the middle than they are at the edges. They work by drawing beams of light towards their center, which is why they are also called converging lenses. The concentration of light beams allows us to see great distances with them. They are thus instruments of unification, bringing what is far away closer to us.

But grinding lenses sends minute particles of glass dust into the air. Unless you're wearing really good protective equipment, which Spinoza evidently was not, the dust can get into the lungs, potentially causing dangerous inflammation. This is what happened to him. So in the process of producing instruments and artefacts—convex lenses, a monistic metaphysics—to help us apprehend a *convergent* world the philosopher is felled by deadly sharp fragments, the *dispersed* elements of that world. It's a melancholy tribute to the perpetual struggle between the search for unity and the seemingly inevitable resurgence of disunity.

Geophysiology

You might not be convinced that the prospects are very bright for discovering how rocks, trees or even some animals might be thinking. I'm open to the idea, but I do understand why it makes others a bit nervous. Maybe it stretches the concept of thought beyond the breaking point. But even if you are not willing to walk down that road with Spinoza, there's another benefit to stressing the importance of complexity the way he does. It is that in addition to foreshadowing the rise of AI and cyborgs, Spinoza is also a proto-ecologist.

The reason for saying this is that he understands the way things are arranged in systems, and that the role of everything in the system in which it is placed determines what that thing actually is. This insight foreshadows 'geophysiology,' the study of the Earth qua superorganism. In this section, we'll take a brief look at the basic features of this concept, show how Spinoza is really the first geophysiologist and finally explain why this is all so important from the standpoint of the climate crisis.

Here is one the 20th-century's most provocative geophysiologists, the British polymath James Lovelock, describing the concept of cybernetics, geophysiology's theoretical forerunner. According to Lovelock, cybernetics as a discipline is about understanding "control systems":

> Information is an inherent and essential part of control systems. . . . Whether we are considering a simple electric oven, a chain of retail shops monitored by a computer, a sleeping cat, an ecosystem, or Gaia herself, so long as we are considering something which is adaptive, capable of harvesting information and of storing experience and knowledge, then its study is a matter of cybernetics and what is studied can be called a 'system.'
>
> *(2009, 57)*

In a moment, we'll come back to one very special member on this list of control systems, Gaia or the Earth system. For now, I want only to emphasize the extraordinary breadth of the definition. Have you ever noticed that for all their obvious differences a sleeping cat and a chain of retail stores with a central monitoring computer are the same sort of thing? Neither had I, but cybernetics has an amazing allure to it. Once you see the similarities among these kinds of things, you begin to see systems *everywhere*.

So what is the common feature of these things? They are all parts working to maintain the equilibrium of a larger whole. This is pretty vague, so let's consider a simple example, the way your body maintains its internal temperature. If you are exposed to the cold and your core temperature begins to drop as a result, the body will begin to shiver. Shivering generates bodily heat by stimulating muscle activity. When it gets too hot, the body engages in a form of heat dispersal through evaporation (sweating).

However, as Lovelock points out, the goal is not to maintain a single, steady temperature state, what he calls the "mythical" 98.4° F. In fact, the system operates so as to induce the optimum temperature for whatever activity the body is engaged in. When we exercise, it rises well above the norm, and early in the day it falls well below it. In addition, the normal temperature describes only the core, whereas the extremities can function well above or below the norm for extended periods (Lovelock, 2009, 49).

In other words, the temperature regulation system in the human body is attuned to the concrete, and complex, *purposes* or activities of the larger organism it serves. Spinoza too understood this. In a letter to a friend, he once compared humanity, unfavorably, to a worm in the blood. Supposing it could think and have beliefs this worm, says Spinoza, would consider its little world to be the whole world. It would believe that it could explain everything that was going on its immediate surroundings just by pointing to other goings-on in those same surroundings.

But of course the motions of the blood are caused by larger events in the rest of the body of the organism in which the worm resides (Spinoza, 2009, 142). This is a philosophical parable for Spinoza. When it comes to our attempts to formulate as many true beliefs about the world as we can, we tend to be too much like that worm. The worm's chief failing is that it takes what is merely a part to be the whole. If it dared to expand its vision it would discover that the 'whole' of its bloody surroundings is a mere part in a larger whole. The exercise can be repeated as often as our intrepid little creature cares or dares. Because all the world's particulars are connected to one another, in infinite space and eternal time, every newly discovered whole will turn out, on reflection, to be but a part of a still larger whole.

However, it is crucial to emphasize that the ever expanding whole is not blankly uniform. There are two things we could say to the worm. The first would be to encourage it to find a cause for every effect it discovers, converting the cause into an effect with a further cause, and so on. This is a perfectly useful exercise. If my mail is delivered wet, I assume that's because it rained today, which can be explained by the low pressure system in the area, which is itself an effect of warm air ascending into the atmosphere. And so on. But there's different way to look at things. When noticing an event of some kind in the world, don't ask what caused it but rather what *role* it plays in the maintenance of some larger whole.

Your body is a system of systems. Each thing gets to self-regulate more or less in its own way, making it distinct from, but also connected to, the thing just next to it. Spinoza wants us to imagine all of this complexity while still being die-hard monists. When you think of the world in cybernetic terms monism does not entail monotony. That point is so important it bears repeating: *monism does not entail monotony*. Of course, those who accept some version of the Great Chain of Being might say that *their* picture contains maximal diversity and complexity because it is filled with metaphysically distinct types of things. Trees are different than humans because they occupy distinct spheres of being.

Well I'm not buying it anymore. I look out my window at the Japanese maple in the yard. Just like me it's engaged in an endless process of internal self-regulation. And, again just like me, it's hooked into its surrounding ecosystem from which it draws nutrients and into which it expels wastes. Both of us are individual systems tucked snugly into larger systems. I see different things—tree, human—but not different *kinds* of things. I see unity *and* diversity, or diversity *in* unity.

What's more, me and that tree are not only engaged in similar, though separate, metabolic processes. We are also connected parts of the Earth system. We're not just doing the same kind of thing in our distinct ways, we're doing it *together*. I'm expelling carbon-dioxide for it to breathe even as it is expelling oxygen for me to breathe. If you extend this vision out far enough you eventually arrive at the notion that the Earth is a kind of organism comprised of a dazzling array of interconnected subsystems. This is geophysiology.

To elaborate, we might ponder the workings of just one of these subsystems, the carbon cycle, an obvious focal point for any analysis of the climate crisis. According to Earth system scientist Tyler Volk, the famous Keeling Curve, based on data gathered at Mauna Loa observatory in Hawaii, "shows geophysiology in action, the alternating dominance of photosynthesizers and respirers on the grandest of scales" (2003, 9). The graph displays the bi-annual flux of CO_2 in the northern hemisphere from terrestrial and marine reservoirs to the atmospheric reservoir and back again. The exchange is about 7 ppm in each direction, over and over again. Gaia's very breath.

Think about this on a temporal scale way beyond that captured by the Keeling Curve. As Lovelock points out, there has been life on Earth for about 3.5 billion years. In that time, the sun has been getting progressively hotter, about 25% hotter over that period. In the *really* deep past, the atmosphere contained loads of CO_2 to compensate for the relatively cool sun. But CO_2 levels have come down gradually as the sun has heated up. This did not happen on Venus, which has high and more or less unchanging levels of CO_2.

Why the difference between the two planets? The short answer is that on Earth life *itself* has regulated the amount of CO_2 in the atmosphere in a way that allows it, life, to flourish. Life is regulating the internal temperature of the planet, its body. If it helps, push the analogy with the human body a step further (as Lovelock himself does). Think of the atmosphere as the narrow membrane of the living planet, much as a patina of skin envelopes your body's sub-systems. Inside the earthly membrane, everything—humans, trees, oceans, etc.—is just a part of the superorganism that is the Earth. While living for itself, each part is also preserving the whole.

This applies to the climate too, and there's the rub. Now, there's a rogue element in the Earth system, humanity, pushing atmospheric CO_2 to levels not seen in millions of years. This might even make us think we are some kind of disease in the system. How can that budding geophysiologist, Spinoza, help us to understand what has gone wrong here? One plausible explanation of our behaving the way we have been is that, among all the infinite diversity of things on this planet, we've convinced ourselves that we are a superior kind of thing. With that presumption in place, it's easy to show how everything else is but a failed version of us, and can thus either be modified to suit our purposes or expunged. That's been our species' default view since we started thinking about these matters.

Spinoza, however, shows us how we can retain the notion that we're different without assuming that this gives us license to pave or plow over the non-human. Our thoughts are more complex than those of a mollusk, yes, and that's because our bodies-with-brains are more complex than their bodies-without-brains. But do you know whose body is more complex than ours? Gaia's. This is because it contains both our brainy bodies and the bodies of everything else, all of it caught up in the biogeochemical whirls of carbon, nitrogen, phosphorus and oxygen through and across Earth's biotic and abiotic pools. This super-entity can't survive with nothing but *our* puny brains to support it. It needs the complexity it has evolved over the past 3.5 billion years, complexity we are eroding rapidly.

Whereas Descartes could not see beyond human purposes, in Spinoza's thinking there's a distinction between understanding the non-human and dominating it. Just one generation removed from Descartes and in the heyday of dominance-of-nature-by-science rhetoric Spinoza just does not talk this way. This bears emphasizing. One of the world's finest scientific minds almost never speaks of scientific mastery and control. In fact, in the *Ethics* the language of domination is replaced wholesale by that of freedom and knowledge. And knowledge in that book refers to the quest to grasp the whole *as* whole. Even more remarkably, Spinoza characterizes this endless outward push of the mind as a form of *love*. Finally, this all comes with an affirmation that matter can *also* be thought of as a coordinate system. Technosphere and/or beloved object of knowledge. Duck and/or rabbit.

This potential dual epistemic focus means we can resist the push to convert the world into standing-reserve even as we recognize that it is already technologized and will become even more so. We can make the bio-technosphere over in a way that respects and even boosts its diversity. This is the central ethical challenge of the Anthropocene. Spinoza shows us how to think about it coherently because he believes our freedom consists as much in expanding our knowledge of the whole as in bending it to our purposes. He is the very first of the early-moderns to grasp this basic distinction, and that is why grappling with his metaphysics pays such handsome dividends.

Volk argues that geophysiology should be guided by four primary directives: attend to what causes the cycles of matter, find the keystone parts in nature's systems, observe how they are affected by timescales both large and small, and notice the way life and nonlife interact to make it all hum along aeon upon aeon. The directives, says Volk, "provide a set of conceptual tools—a single mental vehicle—for exploring Gaia" (2003, 26). Spinoza would have endorsed this method of inquiry, adding only that it is an exercise driven indissolubly by both love and the quest for truth.

One final question. If Gaia's body is so complex, should it not already *be* thinking with a corresponding level of complexity? Yes, and it *is*, as evidenced by the fact that we now understand both that the Earth is a system and that anthropogenic climate change threatens this system's basic integrity. The myriad voices of ecological warning—those of scientists, novelists and filmmakers, teen school strikers, religiously inspired protectors of Creation, indigenous peoples who have lived with Gaia's rhythms for thousands of years, brightly festooned extinction rebels dancing in the streets of London and Bratislava, and more—are enunciating Gaia's own thoughts, performing her pain. But our political and economic elites are either ignoring them outright or merely paying them lip service, and that is a key aspect of our tragedy.

Conclusion

Come back, in closing, to my weedy lawn. When I look at it now, sitting on my porch with my morning coffee, I feel vindicated. This humble patch of ex-urban land is my little homage to diversity. Don't get me wrong. It's not a celebration of wildness, of something radically independent of the human. Half the weeds growing there have been carried into the region by human-influenced forces and flows. Synthetic fertilizers from neighboring plots blow over on windy days, making my plants grow a little more extravagantly than they otherwise would. Unusually heavy' rains ferry seeds from nearby gardens. It's an untidy Anthropocene garden. But even so, it's not the same as everyone else's. It embodies everything distinctively weird about the Anthropocene. We need to get used to this messiness, well beyond the appearance of our lawns.

I hope you now appreciate the way Spinoza allows us to articulate these matters so well. But as with the other thinkers we have looked at it's an insight that does not come cheap. Spinoza felt the twin pains of persecution and exile as deeply as any of his people, and perhaps more deeply than most of them. In fact, I've suggested, the pain went so deep that he sought to eradicate the very possibility of its recurrence by making reality safe for diversity. This is the wellspring of his effort to take away fundamental metaphysical division while cultivating maximal diversity and complexity in the unified field of life. It's a difficult thought to hold onto, this diversity in fundamental unity. It tends to slip through the fingers like sand. But it's the way we need to think now, if we can grasp it.

However, there are two elements missing from Spinoza's metaphysics. First, nothing in his system *moves*. God or Nature has no history, at least not one that is essential to It. It's as though the world as we find it has always been just this way. In the 18th-century, theories of evolution started picking up steam, culminating in the revolutionary work of Charles Darwin in the 19th-century. Any such account explains things by telling a story about how they developed over time. This kind of historical thinking was not available in the 17th-century—so there's no question of berating Spinoza for neglecting to talk this way—but *we* can't do without it.

Second, although I have been applauding the celebration of diversity in Spinoza's metaphysics, we might be inclined to politicize these insights. Think again of the value of inclusivity. It's all well and good to bring a group inside the circle of moral considerability, but the gesture is somewhat empty if we do not also solidify their status there by granting them the protection of rights. In his political philosophy, Spinoza does argue for a far-reaching ethic of toleration, as I have mentioned, but he does not develop this into a full-blown rights-theory. That came only with the great democratic political revolutions of the 18th-century.

Again, the last thing I want to do is chastise a thinker for not anticipating every important turn in the world of ideas. In the case of Spinoza, this would be especially churlish since he anticipates so much. But we cannot make sense of the demands of the Anthropocene without history and rights. It's time to add these elements to the account we have been developing by examining the final thinker in our historical sequence.

References

Descartes, R. (1985). "The Principles of Philosophy." In *The Philosophical Writings of Descartes*. Cambridge: Cambridge University Press, volume 1.

Goldstein, R. (2006). *Betraying Spinoza: The Renegade Jew Who Gave us Modernity*. New York: Nextbook.

Kenyon, G. (May 9, 2019). "How Weeds Help Fight Climate Change." *BBC Future*. Retrieved from: www.bbc.com/future/story/20190507-weeds-a-surprising-way-to-fight-climate-change. Accessed May 14, 2019.

Lovelock, J. (2009). *Gaia: A New Look at Life on Earth*. Oxford: Oxford University Press.

Nadler, S. (1999). *Spinoza: A Life*. Cambridge: Cambridge University Press.

Spinoza, B. (1994). "The Ethics." In *A Spinoza Reader*, Edwin Curley (ed.). Princeton: Princeton University Press.

———. (2009). "Letter to Henry Oldenburg." In *Modern Philosophy: An Anthology of Primary Sources*. Indianapolis: Hackett Publishing, 137–143.

Volk, T. (2003). *Gaia's Body: Toward a Physiology of Earth*. Cambridge, MA: The MIT Press.

Wollheben, P. (2016). *The Hidden Life of Trees*. Vancouver: Greystone.

8
HEGEL: RIGHTS

Imagine you are a middle manager of a tech firm, in charge of your company's health and environmental policies. One of your tasks is to reduce the company's carbon footprint by 30% over the next five years. You think of yourself as a capable worker and creative thinker, so even though this is a very ambitious goal you believe you can do it, and set out to impress the higher-ups in the company by your diligence and vision. Here's a fairly abstract way to think of three steps you might pass through as you come to appreciate the complexity of your task.

First, you might think of your company and its current health and environmental policies in relative isolation from anything else. The only way to get anything done around here, you think, is by tweaking existing practices, nudging people to perform the behaviors you want them to adopt, maybe capturing some funds from a pool that had been sitting around unused. But you eventually realize that these are all half-measures at best, and that for the really big changes you are planning, more ambitious actions are required.

The next step might be to look outside the company itself, to ponder its position in a larger social and political context. You are pretty sure you could meet your goals if you could find adequate funding, and decide to look for it in the form of private-sector fundraising as well as tapping into government subsidies. But there's a problem. Society and the government are on the whole quite hostile to 'environmentalism' at the moment. You are skeptical that you can obtain the requisite funding just by describing your company's carbon reduction goal to any of the people you need to impress.

Where do things stand now with your project? There's your company on one side and the larger social order on the other. The two seem poles apart. But rather than give up, what if you decide instead that the best way forward here is to try to persuade people that there are synergies they are not seeing? Perhaps you can convince them that if your company achieves its goal, this will have a significant impact on local air quality and everyone's lung health will be improved.

Not only that, but the green initiatives you have in mind could plausibly lead to technological advances that are also good for everyone else. For example, if your company is large enough, its investment in a particular technology might allow another local company, one that manufactures the technology, to finally take their production process to the scale that puts them on the economic map. This could ramify powerfully through the local economy, with significant job growth for the whole region accruing as a result. This possibility could attract the investments you are looking for.

Whether you appreciate the point or not, you have just enacted a philosophically famous method, the dialectic. Stated abstractly, we begin a dialectical process with something conceived as a self-sufficient whole, think of it next in opposition to what is not-it, and finally show that it and what is not-it can form a bigger whole. At this third stage, each is best understood in relation to the other rather than in isolation from it. The *combination* brings out the best in both. Notice that the process of making new wholes need never cease. The new whole that has been created or discovered will tend to find its own other, a brand new not-it, and then hopefully the two can locate some synergies between them. This allows for the emergence of a still-larger whole. And so on, and so on.

The dialectic is Georg Wilhelm Friedrich Hegel's (1770–1831) marvelous brainchild. As I hope the example just offered demonstrates, it's not only a straightforward idea to grasp in its essence, but also describes the way many of us already think and act. It's the key to genuinely cooperative or communal decision making. You can't enter a corporate boardroom or sit on a committee—in a university, an NGO, at a party caucus, etc.—without, at some point, hearing about all the wonderful synergies people have been dreaming up and want to get to work forging. Most people don't realize that Hegel invented creative synergy! On the other hand, he's got nobody but himself to blame for his lack of recognition on this score. For while he probably did not invent turgid and jargon-saturated prose, he certainly perfected this dark art.

Our multi-chapter investigation of famous philosophers in Part II of this book culminates with a look at the Hegelian dialectic. Why? Because it is the ultimate expression of Spinoza's holistic thinking made to evolve historically. Hegel is the West's first truly historical philosopher. He makes everything move and evolve, from plants to peoples. In the process, he gives us a uniquely powerful way of understanding the moment we now occupy, the Anthropocene. As we will see, this is because he conceives of the whole as a system in which political equals come to recognize each other as such, and where this recognition is encoded in enforceable political rights. Moreover, in Hegel's spirit, we can extend the protection of rights well beyond the human sphere to encompass the whole biosphere, as Ecuador has done.

In my view, the Ecuadorian example, which enshrines rights of nature in the national constitution, is a genuinely exciting and inspiring development in our history. Hegel, I'm going to show, gives us an insightful way of comprehending it. But, just as was the case with the other four philosophers considered here, Hegel would not have come up with his metaphysics were he not responding to the key crisis of his time, political revolution. This is where we begin.

The soul of the world on horseback

When we think of the French Revolution certain key events and characters usually spring to mind. One example is the adoption of the title 'National Assembly' by the Third Estate in June, 1789, which set up a government parallel and mostly hostile to that of the French monarchy. Another is the storming of the Bastille in July, 1789, probably *the* iconic event of the Revolution. But there's also the nationalization of church property in November of the same year, a seizure of immense symbolic power in the general effort to curb ecclesiastical power.

On the more grisly side we have the execution, in January, 1793, of Louis XVI, which finally, and quite literally, removed the head of the monarchical serpent. Or, speaking of lost heads, there's always the execution of Marie Antoinette, in October, 1793; or of one of the Revolution's original leaders, Robespierre, in July, 1794. Finally, we might point to Napoleon's *coup d'état* of November, 1799, which made him First Consul, a position he ultimately consolidated to become Emperor. For years afterward, Europe would quake under his rule.

In truth, as even this potted timeline indicates, the French Revolution was an extended period of socio-political disruption, both in France and elsewhere in Europe, lasting from 1789 until at least 1815. Historians are split on the ultimate meaning of these events. Jonathan Israel, for example, argues that they are the historical basis of all the liberal-democratic principles we value today, principles such as a free press, toleration of religion and government by the people (2013, 897–937). Simon Schama (1990) is more inclined to view the Revolution as a mostly uncontrolled episode of deplorable mob behavior, punctuated by acts of horrific violence and violations of the very rights being promulgated by the revolutionaries.

This is, to be sure, a difference of emphasis only. Neither of these eminent historians would fully reject the point being made by the other. The difference of opinion here mirrors the split in attitudes about the Revolution at the time it was happening. There were those, like the English philosopher Edmund Burke, who thought it was an unmitigated disaster. Others, like the German poet Goethe, believed it represented a nearly heaven-sent deliverance of humanity from the shackles of arbitrary religious and political authority. A new dawn.

Again, there's much truth in both assessments. Burke's main point was that the revolutionaries were trying to impose abstract principles of liberty, equality and fraternity on society but that moral principles cannot be imposed from above like that. They are, rather, the organic product of long-established traditions. If you uproot the old values and try to replace them with rootless ones, violence is inevitable. Why? Because nobody really knows how to structure the social and political order in a way that lives up to brand new values. They also don't know how to believe sincerely in them.

A successful revolution requires everyone to be on board. But where is the enthusiasm for its principles supposed to come from if most people have never even heard of these ideals before? The answer is that it comes through intimidation. The Jacobin Saint-Just, Robespierre's henchman and a key member of the Committee for Public Safety (a delicious euphemism), declared that neutrality for the ideals of the revolution was tantamount to rejection of them. He was the architect of the Reign of Terror, which saw over 35,000 people guillotined, eventually including Saint-Just himself as well as Robespierre. As the saying goes, revolutions tend to devour their own children.

On the other hand, Goethe and his ilk were also right: there really was something hopeful and visionary about those principles, something moreover of vital and enduring importance for us as we try to negotiate our way around the strange new political landscape of the Anthropocene. Hegel too was an enthusiastic supporter of the Revolution, but his support was not unqualified (I'll return to this just below). In 1806, Napoleon was at the gates of the Prussian city of Jena (now in the German state of Thuringia), where Hegel was living at this time.

Hegel welcomed the French military's efforts to overthrow what he thought of as a sclerotic Prussian political establishment. In fact, the whole European political order was, he believed, thoroughly broken and antiquated. It was stuck in pre-modern ways of thinking about the relation between the individual and the state. Hegel saw the political world in much the same way Descartes saw the intellectual world more than 150 years earlier. It had utterly failed to enact the principles of moral and political equality championed by the original thinkers of the Enlightenment. The whole point of that seminal event was to cast off what Kant called our "self-imposed tutelage" and live instead as free individuals among other free individuals.

Instead, ancient privileges, nepotism, feudal hierarchies, religious zealotry and economic stagnation were still the rule. Hegel saw Napoleon as the embodiment of the *will* to change all of this, to drag the old European order kicking and screaming into the modern age. Just as important, he thought he had the metaphysics to back Napoleon up, that the little Corsican was the material enactment of his own philosophical principles. To make the point clear, in 1807 he rushed his magnum opus, *The Phenomenology of Spirit*, to publication because he thought this book captured the very spirit of the Revolution. In a letter to a friend he says, "I saw Napoleon, the soul of the world, riding through the town. . . . It is a wonderful sight to see, concentrated in a point, sitting on a horse, an individual who overruns the world and masters it" (Arthur, 1989, 18).

This sounds pretty enthusiastic, so why do I say that Hegel's support for the Revolution was qualified? Because, much like Burke, he believed that abstract principles needed to find roots in civil society, rather than being imposed by violence from above. But, unlike Burke, he *really* liked the principles themselves.

Here, for example, is article 6 of the Revolution's founding document, *The Declaration of the Rights of Man and Citizen*, adopted in August, 1789:

> All citizens, being equal in the eyes of the law, are equally eligible to all dignities and to all public positions and occupations, according to their abilities, and without distinction except that of their virtues and talents.
>
> *(de Lafayette, 1789)*

A republic of virtues and talents in which established economic, religious and political relations hold no privileged legal sway. That's pretty heady stuff, and Hegel was fully on board with it. Many European intellectuals at the time considered the Revolution to be the practical upshot of a philosophy of politics and history that had been brewing for some time. Hegel himself says, "We should not . . . contradict the assertion that the French Revolution received its first impulse from philosophy" (Hegel, 1956, 446).

He had himself chiefly in mind here (also Kant). But why would anyone be smitten by Napoleon, as so many of Hegel's contemporaries were? Everyone in Europe at the time knew that the man was, to understate the point dramatically, a complete thug. In addition to being a brilliant military strategist—probably the finest the world had ever known, especially in his ability to read maps and understand the nuances of topography—he was also a megalomaniac, a pathological liar, a backstabber, a bully and likely even a rapist.

He was mostly indifferent to the fate of his troops. Over the war years, he lost an average of 50,000 of them a year in combat, compared to the 6,000 who perished under the command of his English contemporary and fellow general, the Duke of Wellington. On campaign in Jaffa, Napoleon summarily slaughtered 4,500 prisoners by bayonet, because he did not want to waste ammunition on them. He regularly deserted his army when things turned sour on the battlefield. He did this for instance in Egypt in 1799, making his way surreptitiously back to France, stranding thousands of poorly outfitted soldiers. One of his remaining officers in Egypt, Jean-Baptiste Kléber, noted that Napoleon left the men *"avec ses culottes plein de merde"* (with their pants full of shit) (Johnson, 2006, 56).

Who could valorize such a scoundrel? Hegel is nothing if not a big picture thinker. He is fully aware of Napoleon's, um, shortcomings. But Hegel is so frustrated by the fragmentation of contemporary Europe and by the sheer inertia of essentially feudal political structures that he is willing to countenance this flawed vehicle of destiny in order to bring into being a new world founded on constitutionally protected political and moral equality. In my view, he thinks of Napoleon the man in much the same way as Napoleon thinks of his troops: a mere vehicle for a larger and more beautiful process. He's a flash in the pan, albeit an unusually brilliant one.

What sort of 'big picture' are we talking about here? What view of reality gives a person permission to overlook Napoleon's world-historical callousness and perfidy? Time for the metaphysics.

Experience and history

The true is the whole. Sounds like something Spinoza would say, doesn't it? Actually Hegel said it. Just like Spinoza, Hegel is a metaphysical monist. But although the similarities between the two thinkers are important, it will be even more crucial to appreciate the differences between them. Let's start with the concept of experience. So far in this book, we have been talking about 'experiences' as incidents in the mind that can be isolated from one another. This is a helpful way of trying to figure out what kinds of things they are. That afternoon with your friend and the wine (described in Chapter 6) proceeded this way.

Recall from that discussion the metaphor of the brain-as-factory. When touring this structure we were looking for a single smell, an isolated olfactory event. When our guide pointed to some neuronal firings we were disappointed because *that* did not seem to be the thing we wanted to find. We decided to call these individual events, single elements of the welter of information delivered to us by our senses and emotions, 'experiences.' And remember that this form of philosophical speculation is what leads Descartes down the primrose path of metaphysical dualism. Even Spinoza, though of course he resists the dualism, thinks of experience this way. Were we right to follow this lead? Hegel provides a more comprehensive way to think of experience, one that goes to the heart of his metaphysics and leads us into his famous discussion of masters and slaves.

We'll begin with his novel view of experience. Suppose you have just returned from a lovely vacation with your family. It's natural to sit back and think about this as a pleasant experience. In doing so, you will linger over especially memorable events, glimpses of stunning natural scenery or intense bonding moments with the other members of the family. Maybe you recall isolated sensory experiences, but even if you do they will probably be folded into larger experiential wholes.

Isn't this what we are really referring to when we talk about experience? The fighter pilot has the *experience* of daring low-altitude bombing raids, the concert pianist or rock guitarist has the *experience* of stage fright, the neophyte student has the *experience* of trying to understand Aristotle's metaphysics or advanced calculus. And so on. This strikes me as a very ordinary way of thinking about experience, and as is often the case, there is something important here to say about ordinary modes of discourse. Let's note two things about examples like this.

First, you might not feel inclined to wonder at all about whether the sorts of phenomena we're talking about here are physical or non-physical. You *could* ask this question about them, but it feels wholly irrelevant and pedantic. Who cares if the rich experience of flying bombing raids is or is not a physical thing? In fact, when we put it this way, the question does not even make very much sense. In thinking about the concept of experience in the more expansive way I've been hinting at here we have effectively blocked the path to metaphysical dualism. Not by refuting it directly but by *changing the subject*. That's a distinctly Hegelian move.

Second, if we want to isolate what is really interesting and problematic about the bombing raids, the stage fright or the perplexed student we might decide to connect individual events to larger wholes. The fighter pilot might think about the plight of his victims, the safety of his fellow pilots, how his family will cope with his possible death, the motives of his commanding officers in ordering this raid, the larger geopolitical aims of the war and so on. With a little imagination we can say something analogous about the two other cases. In other words, a properly philosophical analysis of these experiences connects them to larger social and political wholes.

Even more importantly, the new wholes we have come across in this process unfold essentially over time. They have a historical essence or structure. The stressed out musician will be in a better position to appreciate her experiences if she learns how to see them in the larger, evolving context of her own life and career. She might, for instance, see a connection between her stage fright and *past* anxieties about doing class presentations when she was a kid. And then she might recall a certain trick she learned way back then, like focusing on just one person in the audience at a time, to push past her anxiety. Maybe the same thing will work on this much larger stage in front of this much more discriminating audience? She could bolster these insights by thinking about the sort of performer she hopes to become one day, a *future* ideal that might be thwarted if her stage fright is too debilitating.

Again, analogous things could be said about the other examples. What I want to emphasize here is that monism appears to be the default metaphysics for the richer concept of experience we have been describing in our examples. There's just no point in talking about any of this as either purely physical or mysteriously non-physical. It is, rather, richly social, moral, personal, psychological and historical. That *is* its essence. But by saying that all of this belongs under the heading of 'experience' we are implicitly claiming that at the most basic level we are talking about just one kind of thing. Or better, a rich diversity of things which *together* add up to just one thing: experience. This is a parts-whole understanding of reality that brooks no division of things into metaphysically separate worlds.

If I'm right that this broad conception of experience captures our ordinary understanding of the concept, then most of us are already monists. Not only that, we already think of experience in the broadly historical way Hegel stresses. The key is to cease thinking of what goes on in our heads in abstraction from the world we inhabit and which has caused those things to be in our heads in the first place. Like Spinoza, Hegel is not an atheist. But, again like Spinoza, his 'God' is no personal Being sitting in transcendent splendor, spinning out or destroying worlds, judging our sins, while presumably keeping *His* City running smoothly.

Rather, God just *is* the world of particulars. But unlike Spinoza, Hegel's is a world that develops historically. How? Is there a logic to this development? Indeed there is: it evolves in accordance with the three steps of the dialectic outlined at the outset of this chapter. So we need to say more about this structure of historical development. Actually, that initial example—the middle manager searching for synergies—*might* leave us with the wrong impression about how this works. The danger is that it can make the whole thing look a little too conciliatory or harmonious. Hegel's view of history is anything but, so we need to shut down this potential misinterpretation before it gains any more energy. One way to approach the problem is to focus more carefully on that second step in the dialectical process.

This is where the so far self-sufficient thing encounters its other, what we have been calling its 'not-it.' This is the stage of conflict. Why does Hegel emphasize it? Well, ask first why Plato, Augustine and Descartes are metaphysical dualists. One reason, the reason I've been developing in this book, is that they all perceive a deeply *conflicted* world, but because they cannot stand the sight of it they construct another world. Plato's Forms, Augustine's City of God, Descartes' mind-substance: these are all attempts at purification, a concept that came up a lot in previous chapters, most explicitly in the discussion of Descartes. Hegel will have no truck with any such dualistic purification, but he also recognizes that our world is marked by deep conflict.

The solution? Find a way of incorporating conflict *into* your view of the real, don't push it to another world. "History," Hegel tells us, "is the slaughter-bench at which the happiness of peoples, the wisdom of states, and the virtues of individuals have been victimized" (1956, 21). Opposition, strife, even violence: Hegel refuses to turn away from these ordinary features of life and evolution. He *does* have extraordinary faith in the capacity of reason to help us overcome sustained conflict, but he is no facile win-win philosopher.

The reason behind this stance is important. Your so-called resolutions will not be resolutions at all if you have not honestly faced the conflict that precedes them. That just sounds right to me, though we tend to forget the wisdom in it because many of us find conflict uncomfortable and therefore flee it prematurely. In a phrase that captures his view of conflict crisply, Hegel says he will accept no philosophical outlook that "lacks the seriousness, the suffering, the patience and the labor of the negative" (1979, 10).

The labor of the negative. We'll return often to that seminal idea. Of course, history can get stuck at this conflictual level for protracted periods. Remember, Hegel was living through the Napoleonic wars. Various mixtures of allied forces from all over Europe united fully *seven times* against Bonaparte before he was defeated once and for all at the battle of Waterloo in 1815 and exiled to St. Helena for the rest of his life. On the other hand, it would be just as egregious an error to suppose that negativity is all we get in Hegel. He is also not a philosopher of irredeemable tragedy. The best way to appreciate how Hegel encompasses both conflict and reconciliation in his philosophy is to look at his famous master-slave dialectic.

Hegel asks us to imagine the original or founding moment of civil society as a *struggle* between two people. As Hegel scholar Terry Pinkard has argued, the struggle is defined by the demand of each that the other *recognize* his or her normative authority (2017, 24). In the ordinary course of affairs, prior to this originary struggle, people encounter things they can consume according to their desires. That is a primitive way we recognize things that are not us, and force them to recognize us. If I am hungry and kill an animal to feed me and my family, that animal has, in a very crude sense, been forced to recognize my authority.

But when I meet with another human things change. Now I seek recognition from a thing that is also aware of itself as a thing seeking out recognition. The result for Hegel is that the two of us will engage in a life and death struggle. At some point in the struggle, however, one of us will cry 'Uncle!' One of us, that is, will decide that honor is not worth dying for. Now we have a position of fundamental inequality between the two people.

In fact, the victor in this struggle will enslave the loser. At this point, it looks like the master has won the battle for recognition. After all, the slave is by definition forced to obey the master's commands. But things are not so simple, for the master has achieved a merely Pyrrhic victory. The slave obeys only because he fears for his life if he does not. But the master knows that the slave is human, and is therefore capable of giving his assent freely to his projects and actions. Because of this, the master craves something more than mere animal recognition from the slave. He wants the slave to *endorse* acting on his, the master's, commands. But all he gets is a yes–man.

This is not real recognition at the human level and the master therefore remains unsatisfied in his goal of extracting recognition from this other. Victory in the struggle to the death turns out to be hollow. What has gone wrong here?

A mind of your own

The explanation goes much further than the simple failure on the part of the master to see his desires fulfilled. Indeed, most of his desires are fulfilled. That's what the slave is there to accomplish. More deeply, the master has failed in the task or goal of *self*-knowledge or *self*-consciousness. To see why it's right to put the point this way, imagine you are a politician with significant power and an elaborate plan to remake the national health care system. The plan is so bold that you are not even sure yourself whether it is a good idea. In other words, on this issue we can say that you don't quite know your own mind yet. You sort of want to pull out all the stops to create the plan, but . . . also, maybe not, given the potential risks. It might sound a bit strange to describe this as a lack of self-knowledge, but that is in large part exactly what it is.

To be of two minds on some issue, any issue, is to lack a measure of self-awareness. Haven't you ever said or thought, about some difficult choice, 'Oh, I just don't know what I think about that!' It's a telling locution, suggesting that there's something you really do or should think about the issue, but that you can't quite see it clearly yet. In these situations, what should you do to repair your mind? If you're smart, you will seek out other people to help you understand the complexities of your proposal.

But what sort of people? If in the course of this exercise you have a choice between getting the assistance either of a yes-man or of someone with a mind of her own, who would you choose? I hope you would choose the person with a mind of her own, because—other things being equal—she is much likelier to help you refine your proposal in fruitful ways. The other will tell you what you already know, knowledge that is, by hypothesis, inadequate. One will enhance your self-knowledge, the other will leave it mired in its current rut.

Now generalize the point and observe how far-reaching its implications are. Hegel is insisting that bona fide self-knowledge is a fundamentally social process. Each of us comes to know who we are only through interaction—both discursive and non-discursive—with other people. But for this process to work the people involved in it must be something more than slaves, there for nothing but the satisfaction of our desires. We've arrived at something of a paradox, then. To have a mind of your own requires developing that mind *with others* who also have minds of their own. That paradox goes to the heart of Hegel's social and political philosophy.

If the master/slave dialectic is a parable, what larger moral is Hegel trying to impart with it? Can we go beyond the analysis of how individual minds are made and say something of import about whole societies or even the Earth system? Yes we can, because the quest for ever-increasing self-consciousness is at the heart of all reality. Recall that for Hegel the true is the whole. The *Phenomenology of Spirit* is constructed as an extended elaboration of that idea, so let's pursue it a little further. To avoid any kind of dualistic trap, Hegel insists on thinking of universal history as a process of increasingly sophisticated and extended moments of consciousness. We know who we are only insofar as we come to appreciate the larger wholes in which we are embedded.

Whose consciousness counts here? Most obviously, that of all humans. That's an idea Hegel took directly from the spirit of the Revolution and never really abandoned. However, the picture Hegel is drawing is compatible with our discovering that other *kinds* of beings are a part of this process. The capacity for recognition is all that counts. In principle, this capacity could extend both 'downward' and 'upward' from humanity. That is, it could extend both to some non-human animals—or even to ecosystems—and to cyborg or artificial intelligences. Whether or not such things are capable of mutual recognition is, in other words, an entirely empirical question. If they are, then they can be brought into the process of increasing self-consciousness.

What about Gaia, the entire Earth system? Well, why not? If all individual living things are parts of the superorganism that is the Earth, then the planet itself can be a thing that 'demands' recognition of just this fact about it. But for this to be true it would have to be the sort of thing that has vital interests and that can, in some sense, come to know itself. Is it that sort of thing? At the end of the previous chapter, I offered a way of thinking about this. It was to suggest that *humanity* is the thinking part of the Earth system.

Hegel provides a way to conceptualize this idea that goes beyond what we can glean from Spinoza. This is because with the philosophical tools Hegel invented we can think of the gradual development of Earth system science as Gaia coming to know *itself*. What we almost invariably say in summing up our knowledge in this area is that we are learning more and more about the Earth system. Very true, but if it is as all-encompassing a system as our discoveries suggest, then it is equally true to say that the Earth system is coming to know itself *through* us.

What is this process of or about? This is equivalent to asking what its purpose or goal is. Where is it going? The answer is, it has *no* goal or purpose external to its own movement. This is a vital point, because if we did posit an external goal or purpose we would be slipping back into some form of metaphysical dualism. Think about it. We might say something like, history is the unfolding of God's plan for humanity. But that means there's a God external to this unfolding, with plans and a realm all Its own. Just like that, reality has been bifurcated into two fundamental arenas, what goes on down here and what goes on up there.

The problem recurs even with a secular story. We might say that the goal of history is extending the Pax Americana, the state of peace and security that has allegedly reigned over the world since the end of the Cold War with the rise of the essentially benign American hegemon. Really? Whoever believes that is also dividing the world up, in this case into America and its allies on the one hand and everyone else on the other. The endless War on Terror, which petrifies an Us and Them mentality, gets its juice from the myth of the essentially benign Pax Americana. There's probably no better literary representation of this delusional thinking than Graham Greene's *The Quiet American*, aptly described by one critic as a tale of "America's love affair with its own innocence" (Rieff, 2020).

In both these cases, the meaning of history has been *imposed* on it by force or decree. If you buy either story (and of course a disturbing number of people believe *both*: the Pax Americana *is* God's plan for humanity), you can say quite a bit about how history will go in the future. You will naturally adopt the mantle of the prophet. Hegel's view is very different. He believes that philosophy has no business prophesying. The most it can do is identify the motor of history, the dialectic, and then say how it was operative in past events. It is always and only retrospective. And being specific about things at this level is fully compatible with saying that the future is open. The dialectic often proceeds at a frustratingly creeping pace, with lots of hesitation and backsliding, and when it does produce new wholes they often come in shapes we could not have imagined beforehand. History is about nothing other than itself.

Alright, but saying that history is about nothing but itself leaves open the possibility that there is something to reality other than history, doesn't it? If I say that beer contains nothing but water and hops it doesn't follow that everything is beer. To the likely chagrin of American Supreme Court Justice Brett Kavanaugh, there are lots of non-beery parts of the universe. Hegel (who was also very fond of beer, by the way) shows just how radical a thinker he is by denying that there's anything to reality besides history. That is, he thinks of the whole—the *whole*

whole, if you like—as nothing but consciousness coming to awareness of itself *as* the whole, as what it really is. There is literally *nothing* behind, beyond or beneath this historical process. It is everything.

Another way to put the point is to say there is no single future pulling the present towards it. That's what makes prophecy a misguided enterprise, according to Hegel. At any present time, there are multiple possible paths for the dialectic to take. This is *not* to say that all such possibilities are on a par, as far as the probability of their future occurrence is concerned. Some are obviously more likely than others given the array of path dependencies that define any present moment. From the Hegelian perspective the IPCC is not wrong to lay out specific future scenarios regarding climate impacts, each scenario tied to policies we adopt, or don't adopt, now.

As we have seen, we are likely in for a very rough ride in the early decades of the Anthropocene, but we can still influence the shape the epoch ultimately takes. What I hope for is an approach to this shaping process that stresses and enhances our understanding that we are Gaia thinking itself. This could have significant implications for how we decide to care for this whole. What implications? Above all, the recognition that the Earth system is a thing that qualifies for rights-protection.

The evolution of freedom

For Hegel, rights and freedom are closely linked concepts. This is easy to appreciate. I can ask the abstract question about whether or not I have free will, but when most of us think about freedom we mean it differently. We believe we are free to the extent that our ability to get on with our lives by our own lights is protected by a legally enforceable catalogue of rights. Not that we should be able to do whatever we want full stop, but that our doing whatever we want is constrained only by the rights of others to do the same.

For Hegel, history is the evolving expansion of freedom thought of in roughly this way. The process has three key phases, found respectively in the Orient, the Greco-Roman world and contemporary Europe. In this section, we will look briefly at each of these, but before getting started on that a quick disclaimer is in order. Hegel is a child of his times and when it comes to knowledge of other cultures, including those in the past, his times were not especially enlightened. He writes from a deeply ethnocentric perspective.

Pinkard is right on the money in castigating the way Hegel understands these peoples, an understanding that essentially reduces them to "failed Europeans" (2017, 50–68). Hegel views these cultures not as centers of civilization in their own right, with rich political and artistic heritages, but as political entities defined by the extent to which they fall short of the European ideal. That is something genuinely to deplore, but it should not dissuade us from mining Hegel's *Philosophy of History* for the nuggets of wisdom it contains. Instead, we should approach this book as a kind of historical myth designed to illuminate *our* world more than it does the worlds of the past. It is telling us something about what is of most significance to us, normatively and politically speaking, as moderns. How does the story go?

The narrative of evolving freedom is disarmingly simple. It begins in the Orient, which for Hegel encompasses China mostly, but also ancient Egypt, Persia and India. In these cultures, freedom is concentrated politically in one person, the Emperor or other leader. The people here are basically children, and religion and morality tend to be steeped in mysticism. What Hegel is trying to stress with this characterization is that with these cultures there is a kind of unity, but it is a simple unity because it is internally undifferentiated. There is the leader, whose awful authority is shrouded in a mystical haze, and the mass of people. It is a very primitive state of affairs, but freedom has at least appeared on the historical stage.

The next phase belongs to the world of the Greeks and Romans. The most significant fact about these cultures, for Hegel, is that they were based on an economy of slavery. In other words while in the Orient one was free, here some are free. The advance over the previous culture can also be stated in terms of internal complexity. Generally, the more units or nodes of freedom there are in a society, the more free people there are in it, the more complexity there is in it. This is because of what freedom is.

Come back to the master/slave dialectic. If the master comes to realize that he is not getting the recognition he needs out of the slave, then he might free the slave and even seek some sort of communal or political union with him. Now we have a small whole containing two nodes of freedom, in the language we have been using here. Each has a mind of its own but they are nevertheless committed to collaboration on important matters. In other words, this is a whole that is relatively internally differentiated or complex. That is just what is missing—allegedly—in the Orient but emerges in the democratic city-states of ancient Greece.

Even so, those societies found it hard to justify refusing to extend freedom to others, slaves and women most prominently. This is a glaring contradiction in the ancient social imaginary. Working it out historically brings about the third political phase, that which materializes in early-modern Europe. Now the mantra is that not just one or merely some but *all* are free. Of course, even in the Europe of the early 19th-century this was not true. To take the most obvious point, the slave trade was alive and well in many parts of the world. And even where there was no slavery, many political societies, especially in Europe, were decrepit and corrupt monarchies. As we have seen, these were steep social and political hierarchies which perpetuated arbitrary privilege for the aristocracy and economic misery for the masses.

Hegel did not deny any of this, which is why he was so enthusiastic about that figure on horseback at the gates of Jena in 1806. He believed that Napoleon was freedom's warts-and-all avatar. Napoleon, and the revolutionary ideals he brought in his train, would finally bring freedom to all. This would allow for the emergence of a political whole with the maximum possible amount of internal complexity, the optimal number of freedom-nodes. Self-consciousness realized in, by and through the whole.

Again, we're free when we are allowed to go about our business in a way that does not unduly interfere with the freedom of others to do the same thing. The freer a thing is the more able it is, other things being equal, to flourish *on its own terms*. Hegel's twist on this is that freedom so defined has essentially to do with both recognition and reconciliation among the members of a group. Those terms mean exactly what you think they do. We are free just insofar as we take the steps necessary to recognize every other free thing *as* free and to heal the wounds that have so far kept this recognition from taking root in the group.

Is it not refreshing to see a philosopher—especially one as irreducibly difficult to comprehend as Hegel so often is—talk about one of the discipline's central concepts in a way that is so down to earth? To understand freedom as an unfolding project of recognition and reconciliation goes to the heart of what we actually care about in these matters. It contains the insight that the moment of negativity, when an 'it' is in sharpest opposition to what is 'not-it,' comes with real moral, emotional and political costs. Though it has been a necessary stage the contest has done damage to both parties. And the way recognition and reconciliation work is not by having one party lord it over the other, but by incorporating what is best in both into larger and more creative wholes.

For instance, the history of interaction between the Canadian government and its indigenous peoples has been extremely fraught. Over the course of much of the 20th-century, indigenous children were stolen by government and church officials from their families and placed in so-called residential schools. There, they were subjected to a brutal regime of forced assimilation to settler culture. The damage done to them was deep and intergenerational. The current government has embarked on the process of repairing some of this damage. It is a good sign that Canadians have finally renounced this aspect of their past, at least officially. More importantly, they have described the new political project explicitly as one of 'reconciliation.' It may not live up to its full promise, but in my view understanding it this way is a hopeful first step.

The process in Canada has been frustratingly slow, and this points to another important feature of the Hegelian view. The work of reconciliation is always slow for him just *because* it is thick. It is immersed in the sticky, gravity-laden matter of existing moral and political norms. The negative labors because it must work *through* prejudice, ignorance and general social inertia.

This helps explain further Hegel's ambivalent attitude towards both the French Revolution and Napoleon. He genuinely admired the ideals of the revolutionaries, as we have seen, but the Terror was produced by the belief that the process of implementing these ideals could leap entirely over existing social arrangements and the norms supporting them. That's why neutrality about the Revolution could be cast as opposition to it, and everyone became a target. Sweep everything away, all at once! By contrast, remember that Hegel described the labor of the negative as, among other things, 'patient.'

Why is any of this of importance in the new epoch? Because we have arrived at a peculiarly dangerous historical moment. In Chapter 4 I argued that the most significant political opposition of the climate crisis is between epistocratic eco-socialism and oligarchic reaction wedded to ethno-nationalism. In fact, it looks as though we are just entering what could be a prolonged period of hostility between these two camps, a battle for the soul of the new epoch. It would be un-Hegelian of me in the extreme to try and predict where this is going. I have no idea, and neither does anyone else. At the moment, it is in a purely negative, essentially conflictual and even tragic phase and this aspect of it could deepen dramatically as the challenges of adapting to climate disasters put greater strains on basic resources.

Hegel is genuinely helpful here. There's a world of difference between cultivating conflict for its own sake and working through conflict with the intention of moving beyond it. The latter is the path of synthesis and dialectic, the former that of naked power. We must understand that the political conflicts to come will be among rights-bearing individuals, and that this very fact about them should compel us to search for synthesis as we work through our differences. Because rights are not absolute and recognizing some often entails refusing to grant others there will be victims in this process. But this is compatible with asserting that a political process *constrained* by mutual rights-recognition is likely to minimize injustices.

Of course, the other big task has to do with translating these insights into our relation to the non-human world. In my view, because rights are of such normative importance in our cultural self-understanding, we need to press hard on the development of rights for nature. This movement is gaining substantial momentum all over the world. NGOs like the Global Alliance for the Rights of Nature, working with other like-minded groups, have conducted five international tribunals aimed at securing broad protection for natural entities like rivers and ecosystems (International Rights of Nature Tribunal, 2020). Ecuador has now become the first country in the world to enshrine such rights in its constitution. Chapter 7 of that document lays it all out in painstaking detail, asserting the rights to protection of the whole biosphere: the urban ecology, soil, water, ecosystems, etc.

This is admirably strong language. No longer is all of this diversity mere human property. It is instead something with a "right to exist, persist, maintain and regenerate its vital cycles" (Republic of Ecuador, 2011). In the terminology we have been using here, nature has been granted a degree of freedom it never had before. In this place, it can now flourish on its own terms, at least in theory (enforcing these rights has encountered resistance, and there has been backsliding).

A traditional system of rights allows humans to stand as defendants when their rights have been violated, and to seek compensation for whatever damages those violations have caused. Similarly, wherever rights of nature are enshrined ecosystems can be named as defendants if their rights are violated through human rapacity and industrial overreach. And then appropriate compensation for damages can be extracted from the despoilers. This is a model for the whole world. It is Hegelian recognition and reconciliation deployed to protect ecosystems.

I know it is tempting to say that the Anthropocene militates against this way of thinking, since the very idea of an 'age of man' suggests that we are about to *reduce* diversity and complexity. Maybe some people are inclined to think of the new epoch this way, but the reason I have spent so much time with Hegel, and Spinoza before him, is to demonstrate that we are not compelled to interpret things this way.

I'll reiterate the key takeaway from these two chapters, then: *monism does not entail monotony*. Both Spinoza and Hegel deny the alleged entailment explicitly. And the denial is, in both cases, not merely an assertion or a wish. There's a whole metaphysics to justify and explain it. In both cases, the main idea is that we have very good reasons to think of reality, of the whole, as an essentially single, but internally complex, thing.

Hegel thinks he has identified this system's inner logic: the dialectic as an evolving process of mutual recognition culminating in a system of enforceable rights. The rights of nature movement is a shining testament to the power of Hegel's vision. The people behind it have decided to overcome the negative otherness of the natural world, to enfold it into a broader domain of human artefacts—rights too are artefacts—*so that* it may be free to flourish. That's an essentially Anthropocene achievement because it involves both leaving things alone and involving ourselves with them. Or better: leaving them alone *by* involving ourselves intentionally and ethically in their development. In the Anthropocene, we need to learn to live with exactly this sort of paradox.

Together with the already established system of human rights, rights of nature represent the high point of our moral development so far. But this is only a first step. For if we are willing to grant rights to ecosystems, why not also the Earth system, which is really just the system that contains all the planet's ecosystems? This move is, I think, far *less* bold than the decision to extend rights coverage beyond humans in the first place, as Ecuador has done. *That* was the really revolutionary move! The next step is an obvious logical extension of it. Gaia will have achieved full self-consciousness only when it too receives such protection.

Again, from the standpoint of the dialectic the future is relatively open. An Earth system with legally enforceable rights-protection would be a wonderful synthesis. The theoretical architecture is in place to achieve it so long as we are willing to steer the dialectical ship in this direction. For the most part we still view nature as humanity's dialectical other—a 'not-us'—but this is changing. Those pushing hard for rights for nature are thus the new soul of the world.

Conclusion

In talking about rights for non-human things are we not stretching our concepts beyond the breaking point? We all understand what it means to grant rights to humans. And we know how to make sense of the question of why it is important to do so. There's something it is like to *be* a human being, after all. This is a product of the fact that we all live our lives, as it were, from the inside. And almost all of us want to experience life more or less on our own terms, without arbitrary interference from other people. We want to live life according to our own purposes and reasons, not have these imposed on us by others against our will. In short, things can go well or badly for us and rights go some distance towards helping us make our lives go well.

But there's nothing it is like to *be* a tree, an ecosystem or Gaia, is there? These things don't possess an internal life the way we do. If this is correct then surely it follows that things can't go well or badly for them. And if *this* is the case, then we might think it inappropriate to grant them rights. The Ecuadorians, we might think, are confused about these matters. However, we need to push back on a key inference made in this argument. The claim is that because they lack an internal life, complete with some degree of conscious self-awareness, there's no sense in which things can go well or badly for trees, ecosystems and Gaia.

Why do things have to be exactly *like* us to warrant legally enforceable protection *from* us? As mentioned in Chapter 5, Brazilian leader Jair Bolsonaro has set in motion plans to deforest the Amazon on a massive scale so as to make way for ranchland. At the time of writing (summer, 2019) huge swaths of the forest are ablaze, fires set intentionally under the permissive eye of Bolsonaro and his cronies. This is *ecocide* plain and simple, and many believe the UN should find a way to criminalize it and charge members of the Brazilian government accordingly. If it makes sense to talk this way in the case of Brazil, then the point should generalize.

Wherever we find clear cases of injury to the biosphere we can in principle name the responsible party and go on to seek punishment of and compensation from them (Stone, 2010). But we won't take this way of thinking seriously unless we stop treating all non-human living entities as more or less imperfect versions of ourselves. The focus in this and the previous chapter on protecting and enhancing the world's diversity—both human and non-human—is meant to help get us out of this conceptual rut.

Where do we go from here? Again, philosophy is not in the business of prophecy, so that question cannot be answered with very much granularity by a philosopher. We can, however, say something from this disciplinary perspective about the future's broad contours. In addition to focusing on expanding the rights-revolution what we have learned from Hegel is how important it is for us to think of the human enterprise historically, to plot ourselves on some kind of timeline and try to makes sense of things in terms of that construct.

But what kind of timeline is appropriate for us? Even if we think of the last 300 years or so as a time of technological and moral progress—with lots of backsliding, especially on the moral front—nobody should doubt that the Anthropocene is going to pose severe challenges to our ability to keep the trend moving in this broad direction (the impacts of COVID-19 are a clear harbinger of the challenges to come). We rightly want to know more about what it is going to take to do just this and how, in particular, it is related to the development of the technosphere.

So we need to explain two things. First, can we construct a picture of history that is mostly forward-moving, but that also loops back on itself as crises emerge? What, in general terms, would that picture look like? Second, what role is technology going to play in maintaining that broad forward movement? And how is technology related to our newly emerging, fully *monistic* historical self-definition? Our final chapter takes up these two questions.

References

Arthur, C. (1989). "Hegel and the French Revolution." *Radical Philosophy* 52, 18–21.

de Lafayette, M. (1789). *Declaration of the Rights of Man and Citizen.* Retrieved from: https://www1.curriculum.edu.au/ddunits/downloads/pdf/dec_of_rights.pdf. Accessed October 17, 2018.

Hegel, G.W.F. (1956). *The Philosophy of World History.* New York: Dover.

———. (1979). *Phenomenology of Spirit.* Oxford: Oxford University Press.

International Rights of Nature Tribunal. (2020). Retrieved from: www.rightsofnaturetribunal.com. Accessed February 19, 2020.

Israel, J.I. (2013). *Democratic Enlightenment: Philosophy, Revolution and Human Rights: 1750–1790.* Oxford: Oxford University Press.

Johnson, P. (2006). *Napoleon: A Life.* New York: Penguin.

Pinkard, T. (2017). *Does History Make Sense? Hegel on the Historical Shapes of Justice.* Cambridge, MA: Harvard University Press.

Republic of Ecuador. (2011). *Constitution of 2008 (Updated).* Retrieved from: http://pdba.georgetown.edu/Constitutions/Ecuador/english08.html. Accessed April 29, 2019.

Rieff, D. (February 15, 2020). "Powered Out: Samantha Power Misunderstood Her Role." *The National Interest.* Retrieved from: https://nationalinterest.org/feature/powered-out-samantha-power-misunderstood-her-role-123311?page=0%2C3&fbclid=IwAR2tO5aXc5Lvx6H3NG_2FTlEMTJhKw9pYfsdlP4RhwoIcOVXzubgAqZwku8. Accessed February 19, 2020.

Schama, S. (1990). *Citizens: A Chronicle of the French Revolution.* New York: Knopf.

Stone, C. (2010). *Should Tress Have Standing?: Towards Legal Rights for Natural Objects.* Oxford: Oxford University Press.

PART 3

Reorientation

9

ANTHROPOCENE MONISM

Just up the road from my hometown in the suburbs of Vancouver is one of the world's best ski resorts, Whistler-Blackcomb. The oldest part of this mountain complex—Whistler proper—is capped and fenced in by powder-packed bowls of forbidding steepness, unmarked rock cliffs dotted all over them. Learning how to master this terrain is exceedingly difficult. It feels as though gravity is aching to hurl you downward. Because you are naturally disinclined to smash your skull on the rocks, in the early stages of the process the urge is always to fight this force by overcompensating in the other direction. You seek the refuge of the snow-wall behind you and as a result lean too far back on your skis. This makes you more apt to fall because, given the steepness, your skis tend to fly out from underneath you.

The correction? Inch your body gradually forward, get *over* your skis, but not too far. This is what we generally call 'finding the center of gravity.' It's a way of cooperating with something that left to its own devices will cause you serious harm. In these bowls, it's scary as hell at first, that subtle forward lean. But once you get it, and then just let it happen, it provides a glorious sensation of unification between your body and the basic forces of the world. That's all there is to skiing these tracts well: finding a sweet spot of *cooperation* with gravity rather than fighting it all the way down. The feeling is about as close to a state of grace as you are likely to experience, at least if you're an atheist.

We can think of technology this way too. In 1951, the Canadian philosopher of media, Marshall McLuhan (1911–1980), wrote a book called, *The Mechanical Bride: Folklore of Industrial Man*. It's a look at the way print media and advertising convey their messages not so much through the content of the stories or the ads as by the way they are displayed on the page or the screen. In the book's Preface, McLuhan cites a story by Edgar Allen Poe, 'A Descent into the Maelstrom.' It's a tale about a sailor cast from his ship in a storm. A great whirlpool had formed in the middle of the sea and was sucking everything into its destructive gyre.

Eventually trapped in it too, and hanging onto a barrel, the sailor decides not to try and fight the swirling current.

Summarizing the importance of this allegory for coming to a proper understanding of media in our lives, McLuhan says:

> Poe's sailor saved himself by studying the action of the whirlpool and by cooperating with it. The present book . . . [attempts] to set the reader at the centre of the revolving picture created by these affairs where he may observe the action that is in progress and in which everybody is involved. From the analysis of that action, it is hoped, many individual strategies may suggest themselves.
>
> *(1951, v.)*

McLuhan goes on to say some things that are not very useful for our predicament. For example, the sailor is described as finding his way out of the maelstrom by adopting the attitude of an amused spectator. McLuhan will adopt a similar attitude in presenting his media-saturated tableau of current affairs to his readers. A little amusement can't hurt, but we don't have the luxury of being mere spectators to our affairs. That aside, the message is right.

In this chapter we will explore a new metaphysics for the climate crisis, what I call Anthropocene monism. This is the idea that there's now a single, metaphysically basic thing: the historically evolved and evolving technosphere. As this description of the view indicates, it has two key elements. First, a philosophy of history. Hegel has set us up for this but our age requires a philosophy of history all its own, a synthesis of two broad philosophical patterns: the recurring circle and the progressive line. On the one hand, we are children of the Enlightenment and as such cannot altogether renounce the notion that history moves in a linear way from worse to better. But on the other hand, the climate crisis complicates this picture immensely. In particular, it means that history's progressive arc will periodically stop or stall, rolling back on itself in more or less trying cycles of adaptation to catastrophe.

The second element of Anthropocene monism involves looking at technology in the spirit of Poe's sailor. What if instead of bemoaning and fighting our urge to build, we cooperated with it? Our crisis has been caused by technology, but technology poorly guided. Because we have allowed the powerful to convince us that technology is both autonomous and value-neutral, we have ceded control of it to them. So the recommendation to cooperate with technology's momentum is sound *only if* accompanied by the resolution to take political control of the technosphere's design. In the chapter's second half we'll explore the intricacies of this claim, but we begin with the philosophy of history.

Is history cyclical?

A basic contrast has defined historical thinking since philosophers and theologians began constructing grand narratives about our place in the whole. It's easy to state in the abstract. Some have thought of history as the ceaseless repetition of cycles, while others see each event as a specific, and unrepeatable, moment in a linear sequence. In this and the next section of the chapter we'll examine these models in turn, starting with the circle. In the classical Greek and Roman framework, history was almost always viewed in cyclical terms. These accounts were invariably caught up in a larger mythological and poetic way of understanding the world.

Despite being a common piece of the *ancient* worldview, no philosopher ever expressed the cyclical view better than the thoroughly *modern* philosopher, Nietzsche. He describes living with the insight into historical repetition—what he calls the eternal return of the same—as "the greatest weight":

> What if some day or night a demon were to steal after you . . . and say to you: 'This life, as you now live it and have lived it, you will have to live once more and innumerable times more; and there will be nothing new in it, but every pain and every joy and every thought and sigh . . . must return to you—all in the same sequence.'
>
> *(1974, 273)*

Nietzsche did not mean this only as a theory about how history works. It is also, and perhaps primarily, a kind of moral and existential test. He thought the idea would terrify most of us, crush us, nauseate us.

But anyone who had the strength to say yes to life, to not only not commit suicide but exist joyfully while in the grip of the insight into the eternal return of the same, was his kind of person. However, the idea *is* clearly also about the structure of historical development. If you endorse it at the ethical or existential level you are committed to having a specific opinion about what history *means* or how it works. You are trying to locate your life in a larger temporal whole. Such literal cyclical repetition, in one form or another, would have been endorsed by just about every ancient historian or philosopher.

One needn't insist on the literal repetition of events, as Nietzsche does. Instead, the claim might be that *structures* repeat themselves endlessly, but that the content of the structures is different from one historical epoch to another. The early-modern Italian philosopher Giambattista Vico (1668–1744), for example, held that history is a story of the recurrent rise and fall of civilizations. What is remarkable about Vico is that he worked out this structure in minute detail for whole historical epochs.

The structure itself is quite simple: "Men first feel necessity, then look for utility, next attend to comfort, still later amuse themselves with pleasure, thence grow dissolute in luxury, and finally go mad and waste their substance" (quoted in Löwith, 1949, 133). Sometimes the final stage, the descent to madness, can be deferred if a people manages to empower an iron-fisted monarch from within or is conquered from without. But in the absence of either of these "remedies," there will inevitably be a reversion to barbarism, where brute "necessity" governs human lives. The term Vico uses for this last outcome is "recourse."

Theories like this issue from a deep fascination with, as well as fear of, nature. Our most basic understanding of cyclical processes comes from what we see going on in the natural world. We watch the seasons turn over and over, the endless and bloody dance of predator and prey, birth, decay and death. Are we not right to wonder at, and be terrified by, the sheer persistence of these cycles? Not only do they never cease, but our designs and purposes are fully at their mercy.

Our technologies, from vaccines to sea walls, are meant to shield us from their power, and these artefacts *can* provide shelter and protection for a time. Depending on how well made they are, they allow us to forestall an otherwise early and inglorious death at the hands of some entity or natural force that is indifferent to our existence, like COVID-19 or rising seas. But the forces in question are relentless. They inevitably bring down our walls, our cities, our governments, our very lives. This, we might think, is simply how things must be.

It is possible to derive a deep and abiding humility from this worldview. The 20th-century German philosopher Karl Löwith (1897–1973) captures this attitude nicely in describing the philosophical lessons the ancient historian Polybius seeks to impart to his readers, most of whom were already intuitively convinced of the cyclical view of history:

> The moral lesson to be drawn from the historical experience of alternating glories and disasters is, according to Polybius, 'never to boast unduly of achievements' by being overbearing and merciless but rather to reflect on the opposite extremity of fortune.
>
> *(Löwith, 1949, 9)*

It is no accident that the ancient Greeks and Romans were so impressed by the power of Fortune to reshape their worlds. When you are attuned to the inevitability of recurrence, when you are convinced that you have little power over the universe's cosmic cycles, you will also be alive to the reality of mutability or change. Your best laid plans will always be laid low by the blows of happenstance.

The ancient myth of Sisyphus is a standing rebuke to those who would defy this reality. Sisyphus is punished by Zeus for chaining up Death, who had come to claim him. He is condemned to the Underworld for all eternity, where he must roll a boulder up a hill only to watch it roll down again. It's a lovely parable. In attempting to stave off death, Sisyphus symbolically defies the power of the cosmic cycle. His punishment therefore is an eternally repeating reminder of the brute fact of *return*. If you fear the gods, the story is a potentially powerful deterrent against the urge to, say, extend life technologically. Best to plod on, accepting that nature will dispense of you and yours on its own schedule.

But then there's Nietzsche, who advises anything but this sort of resignation. Indeed, even in the ancient world, alongside Polybius' recommendation to live with humility and moderation, there arises exactly the opposite sort of response to nature's relentless cycles. Heroic actions were an attempt to put an individual stamp on these monotonous repetitions and so ancient cultures are also marked by the cult of the 'great man' and his (always his) deeds (usually military) (Löwith, 1949, 9). The counsel to humility and the search for glory are probably therefore two faces of the same coin. That is, they are diametrically opposed, but both totally understandable, reactions to the insight into the cyclical nature of reality.

Nietzsche's existential and ethical ideal is heroic. Think of the insight into repetition as the discovery of meaninglessness. That's easy to do. If all our projects will be crushed beneath nature's relentlessly rolling wheel then they are all, in a sense, ultimately meaningless. Don't our projects derive most of their meaning from our projecting them into the future? Don't we want our actions and the values they manifest to last, hopefully well beyond our own lives? Suppose we do and that endless repetition is nevertheless a fact of the universe. In that case we can understand how the *most* heroic people are precisely those who are able to live joyfully and willfully while staring into the swirling abyss.

These ideas about nature's cyclical processes and how we fit into them are not of merely antiquarian interest. In fact, they form part of the backbone of our understanding of ecology. We sometimes think ecological thinking begins in the 1960s. It certainly picks up a good deal of steam at that point but one very important strand of it begins much earlier, in the late 18th- and early 19th—century. In an age that produced a lot of dreary people, the Reverend Thomas Malthus (1766–1834) was an exceptionally dreary figure. Malthus advanced a theory of civilizational rise and decline that has had a lasting impact on the contemporary mind. According to Malthus, there's a basic mismatch between the respective rates at which populations and the means they have to feed themselves increase.

While increases in the means of subsistence rise 'arithmetically,' population increases 'geometrically.' Think of two number series. The first runs, 1, 2, 3, 4, 5, . . . n; the second runs, 1, 2, 4, 8, 16, 32 . . . n. The first series is arithmetical, the second geometric (or exponential as we now call it). You can see that it won't take very long for the geometrical series to outrun the arithmetical one, eventually by orders of magnitude. Malthus claims that this basic function defines the human relation to the natural world, the result of which is a series of cycles. Humans gain a bit of technological mastery over nature, use this to increase food production, then basically go mad having babies until the population is no longer capable of feeding itself.

Much like Vico's recourse, the main remedy for this excess is *not* public welfare (which Malthus vigorously opposed) but natural "misery":

> Sickly seasons, epidemics, pestilence and plague, advance in terrific array, and sweep off their thousands and ten thousands. Should success still be incomplete, gigantic inevitable famine stalks in the rear, and with one mighty blow, levels the population with the food of the world.
>
> *(quoted in Williston, 2015, 323)*

See what I mean by dreary? And in case you were inclined to dismiss this as an outmoded way of thinking, here's an updated, and admirably lucid, version of it from the philosopher Craig Dilworth, who calls it the "vicious circle principle":

> Humankind's development consists in an accelerating movement from situations of scarcity, to technological innovation, to increased resource availability, to increased consumption, to population growth, to resource depletion, to scarcity once again, and so on.
>
> *(2010, 110)*

This kind of thinking is an important feature of modern environmentalism. In 1972, for example, the Club of Rome issued its massively influential *Limits to Growth* report, a dire assessment of the human industrial enterprise. According to the authors of that document our civilization is in full ecological overshoot, and the consequences of failing to constrain ourselves will be Malthusian misery. Jared Diamond's bestseller, *Collapse*, is a gripping narrative of civilizational collapse from Easter Island to contemporary Rwanda, and draws explicitly on this tradition.

Neo-Malthusian views are difficult to dismiss entirely, as their staying power in the contemporary consciousness attests. Perhaps these cycles *are* inevitable, and we should therefore learn to anticipate and even embrace them. Some ancient Stoic philosophers say that spiritual peace—a pleasantly relaxed state of mind they call *ataraxia*—can be gained through the simple contemplation of nature's changeless and fateful cycles (think of Polybius' exhortation). Albert Camus' existentialist classic, *The Myth of Sisyphus*, ends with the line, "one must imagine Sisyphus happy" (1975, 111).

As we have seen, Sisyphus is condemned for all eternity to push a heavy boulder up a hill only to watch it roll down again. No sleep, no checking his email for updates on his stock portfolio, no water breaks. Just that damn boulder. Well, I've given it the old college try and just cannot manage to see how this might be a happy fate. This poor sap's task strikes me as the very quintessence of tedium and futility. There's got to be a better way to think about history.

The synthesis of line and circle

Those who think that history moves in a line form the other half of the contrast we are considering. Hegel is the poster-child here, and since we have already devoted a whole chapter to him we don't need to say too much more about this way of seeing things. The key feature of Enlightenment theories of history like his is their progressivism. This is the notion that science and technology provide the means of making life better, and that all we need to do in order to usher in the new age is unshackle these forces from crippling ecclesiastical oversight.

The typical Enlightenment narrative took the form of a developmental theory. Prior to the enlightened present, we humans have been mere children under what Immanuel Kant called "the self-imposed tutelage" of religion. But now, thanks to modern science, we have thrown off those shackles and achieved a kind of collective maturity (1996, 17). If allowed to develop without religious or political interference, science and technology will produce an entirely earthly paradise. It will make us happy, or happier.

That was a pretty bold thing to say at modernity's dawn, but the Enlightenment is not in every respect as novel as its self-conception makes it out to be. For instance, the linear conception of historical time is an invention of Judeo-Christian eschatology, a branch of theology devoted to the so-called science of final things (death, final judgement, etc.). Both religion and science provide us with what we might call existential shields, ways of fending off suffering and despair by giving us some grounds for hope that the future will be better than the present or the past.

In this respect, they are both designed to help us transcend nature's awful repetitions. In calling them existential shields I mean to highlight the way they are aimed at protecting against the way nature deals out death and suffering. They shield us not by defeating these forces utterly, of course, but by making them *meaningful*. As the philosopher Susan Neiman has argued (2004), to make meaningful in this sense is all about finding a way to live with the inevitable pains and sorrows that beset us as a result of nature's relentless cycles. And those cycles *are* depressing!

Come back for a moment to my dismissal of Camus' take on Sisyphus. I'm inclined to think this worldview is missing something really vital. I'm talking about the ideal of progress. If you embrace this ideal, as I do, then you think that even if it takes a *really* long time for the linear sequence to come to fulfillment, and there are numerous stalls and hesitations along the way, the general trend is upward. But this way of putting the point is potentially misleading. We needn't believe progress *will* happen, only that it is eminently worth fighting for. That it's admirable to act as though things could get better. Of course, we need acceptable criteria of progress. By itself, this concept is evaluatively neutral: it can refer to the spread of liberal-democratic institutions or of the COVID-19 virus. We are interested here in the spread of things we think are positive, like equality and happiness. The point of promoting such values is to usher their negative counterparts, inequality and misery, off the historical stage to the extent we can.

Commitment to progress of this form is difficult to dismiss sincerely, and technology undoubtedly has something to do with what it means. Other things being equal it's hard to promote general happiness when you're battling a deadly virus, and so we need the helping hand of technology in these cases. Recognizing this is all there is to being *enlightened* in the modern age. Still, that historically specific thing, 'the Enlightenment,' definitely presents too sunny a view of the human prospect (Heath, 2014). For we now *know* that technology, wedded to an aggressively expansionist capitalist economic order, has brought ecological catastrophe.

This has transpired because we have lost sight of, or taken for granted, nature's circles and cycles. The Enlightenment was intoxicated with the notion that nature was a thing to be overcome, tamed, mastered, possessed, etc. In response to this hubris the cycles, which we should have learned to understand and respect, have come back to bite us hard. That's what comes of being pathologically committed to linearity. What Kant said about his contemporaries applies even more to us: though we live in the age of enlightenment we are not enlightened.

Discussing the Christian eschatology of the books of Revelations and Daniel, Alison McQueen argues that these representations of transcendent linearity beguile precisely because they have an easy way of explaining pain and suffering away. All you need to do is assert that our travails are a necessary part of history's glorious march to the New Jerusalem:

> Without fear of conflict, evil or ambiguity, the gates of the city will remain forever open. God's final assertion of earthly sovereignty destroys all the boundaries, differences, conflicts, and moral complexity that define the political world. Revelation and Daniel offer their audiences . . . a seductive vision of a world without politics.
>
> *(2018, 42)*

Giddy linearity is, in other words, anti-political to its core. In a brilliant recent analysis, Jedediah Purdy (2018), has said something similar about certain aspects of the American environmental imaginary, for example the frontier ideal. That ideal was a forward-focused tunnel vision, allowing settlers to literally trample over native inhabitants in the course of the westward push.

All of this implies we need a more tempered approach to the concept of historical progress. In fact, something like that is already available implicitly in the relevant literature. Jared Diamond and the Club of Rome are not *really* hard-core neo-Malthusians. They, and we, are too *modern* to believe that history must repeat its dismal cycles endlessly, or that we should seek solace from this worldview. Malthus himself was definitely Malthusian: he welcomed the circle's ravages. By contrast, our so-called neo-Malthusians are saying that we need to change our behaviors or risk disaster. Why would they bother writing these books or issuing these reports if they believed that history were *inevitably* about to turn back on itself like a dog chasing its own tail? More plausibly, they are pleading for a rapprochement between ecology and industry, nature and the economy, the circle and the line.

Here's a way of getting a start on the hoped-for synthesis. Have you ever reflected on the structure of family life? I was thinking about this recently and something I had not explicitly noticed before jumped out at me. It's that family life is a tightly interwoven mix of repetition and forward movement. On the one hand, it is defined by the cycles of diurnal, seasonal and annual activities. Meals together, breaking out and dusting off season-specific tools like your battered snow shovel, that yearly vacation to the cottage or cabin. These are cycles within cycles. On the other hand, there's the kind of development that is mostly non-repetitive. Your kids clear educational hurdles, your careers advance, you downsize, loved ones age and die. Etc.

That's the general structure, then: forward movement embedding cycles—some of them recurrent and predictable, others surprising and possibly painful or disruptive. I don't think family life would make much sense in the absence of either element. It's that notion of sense-making that interests me most here. The family is a unit that typically—when all goes reasonably well—grows in self-understanding over the years. It learns about its own character, its fears and aspirations, and about its relation to the world beyond it, precisely by moving forward *while* looping back at more or less fixed temporal points.

This structure, in turn, exhibits the family's relation to time. The forward movement relates it hopefully to the future, while the backward loops relate it both to the demands of the present and the heritage of the past. We return to *this* place *this* summer because it is now warm enough to go again, and going there is what generations of my family have done in the summer. This is a good way to think about history in general. But here our family analogy breaks down somewhat.

The big problem with it is that the backward circles we are in for will be mostly unpleasant, even tragic. The incursions of these circles into the line's forward movement will stall that movement, sometimes for a very long period of time. These are the moments humanity will be working through crisis, deploying our technological wiles to save what we can of humanity and the biosphere. At such times, we will be forced to take planetary boundaries and moral limits very seriously indeed. Because they will be moments of Hegelian conflict—the labor of the negative—these times will scar us. People will die, species will vanish. However, they will also be opportunities for deeper reflection about where we are going as a species. As such, they can be a catalyst for wider syntheses. What broad form should these moments of reflection on the human career assume?

The answer is that our attention must be fixed on how we can make *this place* better for all of us. If we lose sight altogether of the progressive goal, we will forget this and then get stuck in one or another form of pure negativity. That is likely to be an irredeemably ugly world, a hellscape of moral and political barbarism. It may sound strange to describe it as largely a technological issue, but this is exactly the shape it will assume. Barbarism will involve the perpetuation or intensification of grotesque social and political inequality, which will itself be reflected in our technological research and development priorities. It's a short, sad step from the misallocation of scarce resources in decisions about research to the biased deployment of saving technologies. More and more, we will need technology to save us from cascading crises, but deepening political divisions among us will cause its benefits to be inequitably distributed.

By way of closing this section, let me contrast the progressive, inward design focus with an increasingly prevalent form of outward focus, one that can save only a privileged subset of humanity (if it can save anyone). The efforts of tycoons like Elon Musk and Jeff Bezos to transport us to other planets in the hopes of one day setting up permanent colonies on them, strike me as dangerous escapism. I don't want to rule out this sort of adventure altogether, but I worry both about who gets to go to these inter-stellar wonderlands and how we will govern them once we get there.

Plutarch, the ancient Greek essayist and biographer, reports that when Alexander the Great learned about the philosophical theory of infinite worlds he wept uncontrollably. "Is it not worthy of tears," he said, "that when there are infinitely many worlds, we are not yet masters of one?" (quoted in Rubenstein, 2014, 52). Of course, Alexander is bewailing his incomplete *military* mastery over the known world and that is the last thing I am plumping for. But I concur with the spirit of his lament, that we should knit our world together before looking beyond it.

I would like to hear a novel *political* idea from the boosters of space colonization, but all we get from them is an endless stream of bland and entirely apolitical techno-enthusiasm, shot through with half-baked musings about the 'destiny' of humanity. Given this absence of political and moral seriousness it seems obvious, to me at least, that if our fantasy-engorged plutocrats have their way, we will likely export our demonstrably flawed conceptions of planetary governance to other worlds, eventually visiting on those places the same eco-political disaster we have brought about here. Rather than dreaming up ways for us—a few of us, at least—to leave this place why don't these guys spend their billions on, say, equipping the developing world with technologies for adapting resiliently to climate change?

The obsession with space colonization is of a piece with the ethos of mindless expansion that made these men richer than anyone should be. If growth is a good in itself for Amazon or PayPal, then it must be for the species too. Bezos has said explicitly that the great thing about escape from Earth is that we can expand our population into the trillions, thereby allowing for the proliferation of millions of Einsteins and Mozarts. It is then assumed that a just and ecologically responsible political order will emerge from this state of affairs, like Athena leaping resplendently clad and ready for battle from the head of Zeus. This is an embarrassingly juvenile vision of the future.

When I say we have to embrace our technological present and future I'm talking about enhancing *this* place, and doing so through a robust debate about the way our best values should shape our general attitude to technology, and thereby also our politics. So let's talk about technology.

Who's your overlord?

In my view, there's far too much negativity about technology. I don't mean that people have an overly finger-wagging attitude about it. Some do, lots of others don't. Rather, I'm talking about the sort of negativity Hegel emphasizes, that moment of opposition between *us* and whatever we deem *not-us*. With respect to our technology, there are two senses in which this sort of negativity might manifest. The first is that we might see technology as pitted against nature. Unspoiled nature, on this understanding, is technology's negative 'other.' I won't say much more about that view here because we've already ushered it off the stage (in Chapter 2). It's the second sense of negativity and otherness I will focus on in what remains of this chapter. This is the idea that although we created it technology has become something *other than us*, a force beyond our control. It is our overlord, modern humanity's ultimate not-it.

Although its role in our lives is manifold, ubiquitous and indispensable, theorists of technology do not agree on the exact nature of this phenomenon. In this section, I'll stake a claim in this disputed terrain, mostly so that we have a defensible and working conception of technology. Philosophers have been arguing about the nature of technology for the better part of a century, so as you might imagine things have gotten exceedingly complicated in this area. Our job here is not to examine every detail of these theoretical developments, but to understand more deeply the role technology plays in our lives now and will increasingly play in the times to come.

Two distinctions are key to this analysis (Feenberg, 1999, 9). The first is between technology as autonomous or human-controlled; the second is between technology as value-laden or value-neutral. In what follows in this and the next section we will look at these two distinctions in order and then talk about the ways the key terms might combine to give us a coherent picture of technology's essence.

Is technology something we control or does it develop autonomously, according to its own purposes? At first glance, the question might strike you as rather silly. Technology is a human creation, so of course we are the ones in control of its development. Lurid fantasies of technology run amok—summarized in the ubiquitous Frankenstein trope—are surely just that, fantasies. But the issue is not so clear cut.

An analogy with the market economy might be helpful. In one sense, 'the economy' is obviously a human creation. It's an institution humans have quite consciously set up at a specific point in our history, one whose features we have been tweaking over the decades just as consciously, for instance through various regulatory measures. In principle, so goes the thinking, we could replace the dominant model of free market capitalism with something quite different if we so choose: steady-state capitalism, ecosocialism, communism, etc.

Or could we? Our rhetoric about the economy suggests that we actually think of it as an autonomous force. At the time of the 2008 financial crash, we were bombarded with the message that some of the planet's biggest financial institutions—JP Morgan Chase, Freddy Mac, AIG, Bank of America, Fanny Mae and more—were 'too big to fail.' Because of this, the US government was forced to bail them out financially, to the tune of approximately $700 billion. So who is really calling the shots here, humans—specifically American taxpayers— or this amorphous and quasi-independent thing called 'the economy'? At times like this, it can sure feel as though the latter is an autonomous entity, with aims and purposes all its own. Governments and corporations milk this intuition for all it's worth, informing us regularly of the 'demands' of the economy, to which we must humbly submit. The same sort of thing is now unfolding in the wake of the COVID-19 crisis, though on a much larger financial scale.

But we should not allow ourselves to be swept up uncritically by this way of thinking. The economy is not in fact autonomous. When we describe an institution as too big to fail, and bail it out accordingly, we have made a *choice* about how to distribute the costs and burdens of market failure. This choice is a direct reflection of power relations in society, as is the choice to refuse to regulate financial institutions properly so that crashes like this don't happen in the first place or, indeed, again. Canada, whose banks are better regulated than American ones, did not experience anything like the American financial tsunami, except of course as an *effect* of the American disaster, which the whole world was compelled to endure.

Here's another example. Many people are worried about AI because they foresee a possible future in which a machine superintelligence might evolve from the relatively primitive AI technologies in our world right now. Though it might not be on the horizon, this is not an outlandish possibility. Once AI reaches general intelligence—basically catches up with us—its progress to superintelligence thereafter could be exponentially rapid. Because this entity's intelligence will, by definition, outstrip ours by orders of magnitude, we cannot be confident that the way it decides to develop will coincide with our values and aims.

In other words, it will seemingly have become fully autonomous. The American sci-fi writer Vernor Vinge dubbed this event the technological 'singularity.' As with the economy, the rhetoric of control and submission here is ubiquitous and telling. In 2011, the IBM computer program Watson played against the reigning human *Jeopardy!* champion, Ken Jennings, thumping him soundly. Jennings was unfazed, saying, "I for one welcome our new computer overlords."

The remark betrays a conception of technological autonomy that is extremely widespread in our culture and has been around at least since the Industrial Revolution. The sentiment is captured nicely by the 19th-century English poet, Samuel Butler (1835–1902):

> Day by day, the machines are gaining ground upon us; . . . we are becoming more subservient to them. . . . The upshot is simply a question of time, but that the time will come when the machines will hold the real supremacy over the world and its inhabitants is what no person of a truly philosophic mind can for a moment question.
>
> (quoted in Barrat, 2013, 161)

Arresting sci-fi fantasies aside, however, it's just false that machines "hold the real supremacy" over us. In fact, the phrase is just too vague to make proper sense of. That should give us pause, especially since Butler and many of his contemporaries—those of the "truly philosophic mind" he praises—were dead certain that the age of slavery to the machines was imminent. Those who think a robot or AI takeover is inevitable would do well to notice that the fear first emerges at a particular historical moment, in the heyday of industrialization in England.

Butler was just a child when Friedrich Engels (1820–1895) published his study of proletariat living and working conditions in Victorian Britain, *The Condition of the Working Class in England*, in 1845. Engels paints a truly appalling picture of mass exploitation in the industrial factories of cities like Manchester and Liverpool. Relative to rural areas, disease rates were higher in these places, wages were lower and death came earlier (to both adults and children). Children, as we discover in the novels of Charles Dickens and Elizabeth Gaskell, were made especially miserable. Most spent very little time in school. It was therefore excusable hyperbole to characterize the spinning and weaving machines, or the steam engines, as malign, quasi-purposive beings bent on grinding humans into so much meat.

This is the atmosphere in which Butler made his dire prediction, but the analysis is mistaken. Those who benefit most from a certain technological regime have a clear interest in perpetuating the fable about machine-autonomy. After all, so goes the myth, if the machines are autonomous with respect to the interests of the workers, then they are also autonomous with respect to those of their owners. The machines don't discriminate among us but dictate their terms to *everyone*. This, of course, is also exactly what is going when CEOs and politicians talk about the 'demands' of the economy.

The whole question of alleged machine autonomy reminds me of a debate in 19th-century philosophy of religion. In his *Phenomenology* Hegel had identified the 'Unhappy Consciousness' of Christianity as an example of humanity's self-alienating tendency, our desire to posit a transcendent Being against whose perfection our own capacities and achievements look pretty measly. This basic idea was developed into a full-blown critique of Christianity by Ludwig Feuerbach (1804–1872) and then by Marx.

In Marx's hands, the merely unhappy religious believer becomes the oppressed worker, fed illusions about the afterlife in order to make him or her complacent about the fact of oppression in this life. Summarizing this historical development, philosopher Michael N. Forster argues that the fake promise of other-worldly rewards doled out by the uber-autonomous Maker—or Its earthly deputies—is aimed specifically at downplaying the capacity regular humans have to change their world for the better. It is meant to make them feel impotent and incompetent. These feelings are not resisted because they come packaged with the message that this is just the way things must be. In sum, the religious illusion arises,

> in order to serve the function of reinforcing . . . oppressive social relations by making their negative consequences seem inevitable and tolerable to people, and hence reconciling people to them, so that they submit to them instead of rebelling against them.

(2015, 809)

If you watch the YouTube video of Watson's victory, you'll notice the Microsoft executives and engineers looking on like nervous parents at their kid's first soccer game. When Jennings bows humbly to his new overlords, I can't help imagining those folks thinking he had absorbed exactly the right lesson from his contest with their machine.

Look, I know the guy was just joking, but his joke expresses a deep truth about the myth of machine autonomy: that there are people in positions of power who *want* us to believe the machines have their own agenda. And so I predict we'll hear more in this vein about the various technologies of climate adaptation. This is already starting with geoengineering, portrayed by some as an inevitability now that we have allowed the stock of atmospheric carbon to accumulate so dangerously. The way these technologies will almost certainly reinforce existing geopolitical power disparities is routinely pooh-poohed by the engineers.

Faced with this ubiquitous way of distorting the field of social and political choice, it is imperative to ask who benefits most from any proposal about how to design the technosphere. Whose values are being looked after in the proposal? This brings us to our second distinction, between technology as value-laden and value-neutral.

Guns kill people

Much as we are too ready to think of technology as autonomous, many of us seem overly willing to think of it as value-neutral. There's a connection between the two errors. If nobody is in control of the machines—if they are indeed autonomous—then they are not expressing anyone's values as they go about doing whatever they do. The only possible candidate for value-conferrer here would be the machines themselves, but machines are not capable of conferring value. Only humans do that.

Those who are worried about AI running amok sometimes express their concerns by saying that AI will reconfigure the world according to its values. But carrying out a program—even one that leads to the extinction of humans—is not a value-directed exercise. The reason for this is that, as we saw in Chapter 1, values embody a conception of the good. If I funnel 5% of my monthly income to various charities, then this action expresses one of my values, the one according to which it is *good* to do my small bit fighting AIDS in Africa, reducing childhood poverty in my own community, unseating a morally reprehensible incumbent, or whatever.

No machine is capable of this sort of reasoning, and I have a hard time seeing how it ever could be. In any case, if we reject the claim that technology is autonomous, we *should* also dispense with the notion that it is ever value-neutral. All technology is value-laden. But here's where things get a bit trickier. Sometimes—but *only* sometimes—we can *change* the values that a piece of technology embodies. Admitting this gets us dangerously close to the view that technology is value-neutral, and that *all* of it can be employed to serve ends anywhere on the value-spectrum: from the evil, through the benign, to the beneficial. Is this true? Is all technology fully value-neutral?

To see why it's not, let's talk about gun violence in the US. Peek around the corner of the claim that we can make *any* kind of technology serve distinct values and you will find a smarmy figure leaning against the wall, all too eager to remind you that guns don't kill people, people do. This is no mere fictional character, alas. You can find him all over cable news shows and social media in America, particularly after another group of innocent people has been mowed down by a deranged AR-15-wielding Second Amendment warrior.

Here's a simple scenario to help us come to grips with the issue. There's a handgun on the kitchen table, let's say the wildly popular Ruger Lightweight Compact Pistol (LCP), the Honey Nut Cheerios of house guns in America. It can be picked up and used by someone bent on harming innocents or a person who decides to tuck it away somewhere safe and store the bullets in another place, to be taken out only in the unlikely event that it is required to defend kith and kin. These two functions embody distinct and opposed values: the destruction of innocents versus their protection. The gun itself, sitting there inertly on the table, does not *contain* either value. So the gun is value-neutral, right? This sounds eminently reasonable, but it's a mistake.

We are, let's say, talking about a kitchen table in the suburban home of an ordinary family in Knoxville, Tennessee. Let's ask a really basic question: why is there a gun in the house at all? The answer, of course, is that this hypothetical family lives in a society that has *decided* to make legal civilian possession of firearms a cornerstone of their democracy. The right to bear arms, as laid out in the Second Amendment of the United States Constitution, "shall not be infringed." This decision expresses a *value* that many Americans believe to be fundamental, that of freedom from potential governmental tyranny (a convenient chimera, but let's not go there). And remember, values embody conceptions of the good. The idea must therefore be that it is a *good thing* for America to be awash in deadly firearms. Since tens of thousands of Americans are killed every year in gun violence, this value clearly overrides that of public safety.

There are other values at play here too. *Forbes* estimates that the gun business in America is worth $28 billion (MacBride, 2018). Gun manufacturers are therefore deriving a great deal of economic value from the loftier value expressed in the Second Amendment. So maybe we should be more critical of the idea that any piece of technology floats tranquilly above our values. Let's remember to pan outward from the thing itself, to the people who might deploy it, and from there to the whole culture of which those people are a part.

This move allows us to resist describing *that* gun on *that* table in Knoxville as a value-neutral piece of technology. After all, an analogous household in, say, Malmö, Sweden, will probably not have one on *its* kitchen table. In other words, the real meaning of that single Knoxville gun cannot be grasped except by noticing that it is one among some 400 million civilian firearms in the US. It's the *arsenal* that embeds the specific values identified earlier, values the lone pistol then inherits. So, yes, *in America* guns do indeed kill people, many of them entirely innocent. That's quite literally what they are meant for.

Beyond guns designed explicitly for carnage, there are more ambiguous cases. I think this is true, to varying degrees, of all the technologies explored in Chapter 6. I've tried to be agnostic about all of them, though I have very serious reservations about many of them. Pharmacological moral enhancement, for example, seems like a non-starter. Geoengineering is more difficult to assess. The version of it getting the most press, solar radiation management, has very grave ethical hurdles to surmount before we should approve it. Perhaps paramount among these is the worry about how to govern the use of this technology internationally if it is meant to function as a sort of global thermostat, as I mentioned just above (Williston, 2018, 164–183).

The point is that we should be having open public discussions about which values any of these technologies is meant to protect or enhance. People—not unfettered markets, neo-liberal politicians, corporate lobbyists or engineers—should decide the shape of our technological future. Our critical discussions of all candidate technologies should be focused in the first place on the manifold risks they pose for anyone whose lives will be impacted by their deployment. With respect to the outcomes of those discussions, there are three broad possibilities: we will discover that the technology serves our best values (or that it is entirely benign), that it currently does not but can be changed or redeployed so that it does, or that it obviously does not. In the first case, we should preserve and enhance the technology, in the second repurpose it, and in the third abandon it.

In none of these cases is it correct to say that the technology is value-neutral. Even in the second case, which looks like it might be analyzed this way, the idea is that the technology currently embodies undesirable values but can be repurposed to embody better ones. That's subtly, though importantly, distinct from the false claim that all technology floats free of our values. Clarifying which values are served by any piece of technology is thus a way of asserting our autonomy over its development. And because *public* toil and treasure are required to develop all technologies, controlling the technological agenda is an essentially political enterprise.

Let me close this section by linking these reflections about technology to the ones from earlier in the chapter about history's shape. While we're marveling over our new-found autonomy over our technologies let's please avoid mindless Prometheanism. I don't want to open the door here to naïve linearity, an approach to history and technology that, as we have seen, too readily becomes anti-political. As William E. Connolly notes, the cyclical way of seeing things can blunt this dimension of linearity. Taken to heart it can encourage us to,

> doubt the providential image of time, reject the compensatory idea that humans can master all the forces that impinge upon life, [and] strive to cultivate wisdom about a world that is neither designed for our benefit nor plastic enough to be putty in our hands.
>
> *(quoted in McQueen, 2018, 197)*

So: some humility to temper our new sense of autonomous technological purpose. Still, the tragic attitude too can be taken too far. One problem with it, at least according to McQueen, is that the humility it recommends can result in quietism, which functions as a passive warrant for the status quo (2018, 199). This is exactly why the cyclical view must itself be pushed out of its comfort zone by reminders that we can and should work to improve our world, to create historically novel social formations and to develop genuinely exciting and emancipatory new technologies.

Here then is the new normal: circles embedded in the line; a complex, historically evolving dance energized by decisions about how to design the technosphere fairly. The signature psychic paradox of this time is that we must build, live and love in this world with both humility and confidence. As a form of metaphysical monism, the picture I'm urging on you allows for no world above, beyond or behind this climate-ravaged planet of ours. It's one thing, made of and by technology and moving historically according to the broad contours laid out in this chapter. All such movement will have crisis as its backdrop. Assuming collective self-deception is not an option there's no dualistic escape hatch for us, as there was in the past.

I have tried to bring this historically specific reality into conceptual focus by articulating it at the level of a metaphysics. This, finally, is what I offer as an alternative to blank anxiety and bewilderment: a moral clarity made sharp and gritty through perpetual moral and political struggle. If the task sounds daunting or pointless bear in mind that saving even *some* of what is valuable is a victory if the only alternative is pure loss. In any case this is the world we've made. Maybe one day we'll find our way to something more purely affirmative and heroic, but for now this strikes me as the most honest worldview available to us. Let's internalize the metaphysics appropriate to it—Anthropocene monism—and see if we can remake the rest of the shared world in the light it casts.

Conclusion

My goal in this this book has been to illuminate what it means to be human in this strange time. That's all philosophy can do, but even so it's a way of focusing the mind that is uniquely capable of providing new orientation to our lives. And because it refreshes our sense of how the human enterprise fits into larger wholes this exercise can alter the entire social imaginary. I have spent so much time on the history of philosophy in Part II because I believe that, done with imagination, metaphysics can bring about comprehensive social, political and personal change. It *has* done this in the past.

Here's a final reminder of the five intellectual innovations I have isolated. If political power can be constrained by ecological knowledge and love of the whole (respectively, Plato and Augustine), while we strive to protect biospheric complexity (Spinoza) by expanding the rights-revolution to the Earth system (Hegel), we will at least stand a chance of preserving what we value most about our shared world. However, none of this is going to happen—or will be sustainable—if we do not double-down on enhancing the technosphere, which is now effectively the whole world (Descartes). And to have any moral worth, this last step must involve bending all significant future technological development to genuinely democratic ends, as suggested in this chapter.

I hope these ideas have provoked some reflection on how we relate to the rest of the world, and to each other, in the age of climate crisis. Our civilization and the environment are, we might say, *fitted* to one another, much like a peg in an appropriately sized and shaped hole. The fit is not perfect by any means but we would not have survived this long, in such numbers, were it not at least roughly correct. Given our headlong economic development and expansion, especially post-WWII, the fit has become very tight indeed. As we have seen (in Chapter 3), some ecologists and systems thinkers are now talking in terms of our reaching or breaching planetary boundaries. But even though it is simple, the peg-in-a-hole metaphor helps guard against two erroneous, because one-sided, ways of thinking about our historical situation.

The first is to assume that the hole is exactly the right shape for the peg but is still larger, so that we can safely expand the peg—civilization—outwards until it fills the hole, which it may never do. That's the approach sometimes described as "business as usual." It's also embedded in the naïve Enlightenment notion that history moves in a straight, and unbroken, line. No time for those pesky historical circles here. But all the evidence compiled over the past 50 years or so, from the Club of Rome's *Limits to Growth Report* (1972) to the UN's *Global Warming of 1.5° Celsius* (2018), militates against this view. Circles are here to stay. If you doubt this, think of the effects of COVID-19 on all our lives. This is exactly the sort of *break* in linear time represented by all significant civilizational crises. Get used to more events like it.

The other error is to think that history is all circle and no line. It is the belief that the civilizational peg outgrew the environmental hole a long time ago, or was never the right shape for it in the first place. Because of this basic mismatch, as it has over and over again in the past the hole is primed once again to spit out the peg, casting our vaunted civilizational achievements to the four winds in the process. Again, I don't think this view correctly represents the thinking of most so-called neo-Malthusians, but it is not entirely incoherent.

Does anyone take it seriously to heart? Yes. There's an odd group known as the Dark Mountain Project whose Manifesto (2014) leads with this epigraph from American philosopher Ralph Waldo Emerson: "The end of the human race is that it will eventually die of civilization." The Manifesto's authors then go on to argue that only a conscious move to what they call "uncivilization" is desirable. Perhaps they are aware, however dimly, that were this conveniently vague goal ever realized in real policy measures, it might lead to the deaths of billions of people.

The view I'm advocating gets between these two options. It is that the peg and the hole are now fused. They are no longer two things, and I mean this quite literally. But the fusion does not imply historical stasis. The fused thing can, and must, evolve. The event of fusion has had negative knock-on effects in the whole structure, effects that can only be ameliorated by fortifying that structure technologically. As I said in the Introduction, none of this should be construed as utopian dreaming. What I'm talking about is adaptation to ongoing crisis, not a vision of the world beyond crisis. I hope we come out of the COVID-19 experience with an improved perspective on what really matters because we are going to need it long after this virus is defeated. How we respond to the pandemic—socially, politically, existentially—will reveal a lot about what the next handful of decades is going to look like. It is no exaggeration to say that this is a crossroads for humanity.

I'm in no position to offer a detailed blueprint of the wholesale change in perspective we require. That's something we must all figure out together, ideally through the deliberative institutions and mechanisms of a truly democratic or ecosocialist epistocracy. I do believe we can make the world more beautiful, more biologically diverse and more inclusive. But it will inherit none of these qualities if we either let the already powerful determine our techno-political future or retreat timidly from the task of re-designing this beleaguered keep, our one and only home.

References

Barrat, J. (2013). *Our Final Invention: Artificial Intelligence and the End of the Human Era.* New York: Thomas Dunne Books.

Camus, A. (1975). *The Myth of Sisyphus.* New York: Penguin Books.

Dark Mountain Project. (2014). *Manifesto.* Retrieved from: https://dark-mountain.net/about/manifesto/. Accessed March 18, 2020.

Dilworth, C. (2010). *Too Smart for Our Own Good: The Ecological Predicament of Humankind.* Cambridge: Cambridge University Press.

Feenberg, A. (1999). *Questioning Technology.* London: Routledge.

Forster, M.N. (2015). "Ideology." In *The Oxford Handbook of German Philosophy in the Nineteenth Century.* Oxford: Oxford University Press, 806–828.

Heath, J. (2014). *Enlightenment 2.0.* Toronto: Harper-Collins.

Kant, I. (1996). "An Answer to the Question: What Is Enlightenment?" In *The Cambridge Edition of the Works of Immanuel Kant: Practical Philosophy.* Cambridge: Cambridge University Press, 11–22.

Löwith, K. (1949). *Meaning in History.* Chicago: University of Chicago Press.

MacBride, E. (November 25, 2018). "America's Gun Business Is $28B. The Gun Violence Business Is Bigger." *Forbes.* Retrieved from: www.forbes.com/sites/elizabethmacbride/2018/11/25/americas-gun-business-is-28b-the-gun-violence-business-is-bigger/#184ed5c53ae8. Accessed April 1, 2019.

McLuhan, H.M. (1951). *The Mechanical Bride: Folklore of Industrial Man.* New York: The Vanguard Press.

McQueen, A. (2018). *Political Realism in Apocalyptic Times.* Cambridge: Cambridge University Press.

Neiman, S. (2004). *Evil in Modern Thought: An Alternative History of Philosophy.* Princeton: Princeton University Press.

Nietzsche, F. (1974). *The Gay Science*, translated by Walter Kaufmann. New York: Vintage Books.
Purdy, J. (2018). *After Nature: A Politics for the Anthropocene*. Cambridge, MA: Harvard University Press.
Rubenstein, M.J. (2014). *Worlds Without End: The Many Lives of the Multiverse*. New York: Columbia University Press.
Williston, B. (2015). *Environmental Ethics for Canadians*. Don Mills: Oxford University Press.
———. (2018). *The Ethics of Climate Change: An Introduction*. London: Routledge.

INDEX

Printed in the United States
By Bookmasters